COSMIC SECRETS

COSMIC SECRETS

Eleanor Haspel-Portner, Ph.D.

Cosmic Secrets
Eleanor Haspel-Portner, Ph.D.

Copyright © 2017 Noble Sciences, LLC.

All rights reserved.
This book or any portion thereof may not be reproduced or used in any manner whatsoever without the express written permission of the publisher except for the use of brief quotations in a book review.

ISBN: 978-1-931053-14-3

Other titles by Eleanor Haspel-Portner
Astronomy Essentials
Patterns of Orientation
*Marriage in Trouble: A Time of Decisio*n

Author's websites
www.beyondhumandesign.com
www.noblesciences.com
www.consciouslifechoices.com
www.DrEleanorHaspel-Portner.com
www.moptwo.com/DrEleanor

Illustrations created by Eleanor Haspel-Portner, Ph.D

Dedication

*To all seekers of self-knowledge —
this work is for you.*

CONTENTS

ILLUSTRATIONS
1

WISDOM AND SCIENCE: THE EVOLUTION OF NOBLE SCIENCES SACRED SYNTHESIS
5

My Story
7

Noble and the Twins: A Story of Love and Inspiration
9

Similarities between Noble Sciences and Human Design
15

Key Differences between Noble Sciences and Human Design
15

Applying Noble Sciences Sacred Synthesis to Your Life
19

ANCIENT ROOTS: A FOUNDATION OF WISDOM
23

The Tree of Life
23

Hindu Chakras
25

Astrology
27

The I-Ching
29

The Mandala of Synthesis
31

Noble Sciences Charts Explained
32

The Importance of Birth Time
33

The Importance of Birth Location
37

Unlocking Noble Sciences Sacred Synthesis
39

Personal and Universal:
The Field of the Conditioning Climate
39

Step 1: Understanding the Four Layers of Being
42

Step 2: Understanding the Eight Charts
45

Step 3: Understanding the Nine Body Centers
46

Step 4: Understanding the Five Types
60

An Example Chart - Steve Jobs
72

Enhancing Practices
87

Mind Maps
87

Journaling: An Integrative Tool that Enhances your Self
88

POSTSCRIPT
93

Awareness Always Precedes Choice
95

Understand and Transcend
95

Tools
97

Noble Sciences Reports
99

ABOUT THE AUTHOR
Eleanor Haspel-Portner, Ph.D.
101

ACKNOWLEDGMENTS
103

Illustrations

Illustration 1: Dr. Eleanor Haspel-Portner 5
Illustration 2: Noble 9
Illustration 3: Eleanor and Noble 10
Illustration 4: Joe and Ruzle 11
Illustration 5: Litter of Six Kittens 11
Illustration 6: Joe, Ruzle, Bear, Chester, Ted, and Rose 12
Illustration 7: Noble 13
Illustration 8: Mandala of Synthesis 20
Illustration 9: Basic Chart Mandala with Definitions 21
Illustration 10: Tree of Life 24
Illustration 11: Chakras 26
Illustration 12: Color Coded Astrological Wheel with Signs, Planets, and Houses 27
Illustration 13: Annotated 64-Gate Mandala 29
Illustration 14: Yin/Yang Hexagram 29
Illustration 15: Yin/Yang Chart 30
Illustration 16: Hexagram of Heaven-Earth Example 31
Illustration 17: Eight Trigrams Astrological Sign and Name 31
Illustration 18: Chart Mandala 33
Illustration 19: Latitude-Longitude 38
Illustration 20: Developmental Timing 40
Illustration 21: Timing in Birth Chart 43
Illustration 22: Developing Conditioning Field 45
Illustration 23: Body Map with Centers Labeled 47
Illustration 24: Eight Double Trigram Hexagrams 48

Illustration 25: Centers Chart 49

Illustration 26: The Crown Center 51

Illustration 27: The Ajna Center 52

Illustration 28: The Throat Center 52

Illustration 29: The Self Center 53

Illustration 30: The Sacral Center 54

Illustration 31: The Root Center 55

Illustration 32: The Splenic Center 56

Illustration 33: The Axes of Awareness, Three Awareness Centers 57

Illustration 34: The Solar Plexus Center 58

Illustration 35: The Heart Center 59

Illustration 36: The Five Types or Five Ways of Being 61

Illustration 37: Manifestor 62

Illustration 38: Generator 64

Illustration 39: Manifesting Generator 66

Illustration 40: Projector 68

Illustration 41: Reflector 70

Illustration 42: Steve Jobs' Birth Data 72

Illustration 43: Steve Jobs' Prenatal Overview 72

Illustration 44: Steve Jobs' Postnatal Overview 73

Illustration 45: Steve Jobs' Key Worlds 73

Illustration 46: Steve Jobs' Prenatal Solar Charts 78

Illustration 47: Steve Jobs' Prenatal Lunar Charts 79

Illustration 48: Steve Jobs' Postnatal Solar Charts 80

Illustration 49: Steve Jobs' Postnatal Lunar Charts 81

Illustration 50: Steve Jobs' Composite Charts 82

Illustration 51: Steve Jobs' Mandala Prenatal Solar 83

Illustration 53: Steve Jobs' Mandala Prenatal Lunar 84
Illustration 54: Steve Jobs' Mandala Prenatal Lunar Minute 84
Illustration 55: Steve Jobs' Mandala Postnatal Solar 85
Illustration 56: Steve Jobs' Mandala Postnatal Solar Minute......................85
Illustration 57: Steve Jobs' Mandala Postnatal Lunar 86
Illustration 58: Steve Jobs' Mandala Postnatal Lunar Minute 86
Illustration 59: Journaling Mind Map..88
Illustration 60: Color Key 89
Illustration 61: Astrology Legend..90
Illustration 62: Table of Planetary Movement...91
Illustration 63: Hexagram Correspondences to Zero Degree Points
of Each Zodiac Sign 92

Wisdom and Science: The Evolution of Noble Sciences Sacred Synthesis

Noble Sciences is a roadmap to life on earth, providing guidance for events and experiences, that enhance our connection with our Divine Self. It builds on the foundation of the four ancient wisdom traditions including the Tree of Life, the I-Ching, the Chakras, and Astrology. For millennia, these wisdom traditions remained virtually the same, operating independently from one another. Each contains information that can give aspects of truth on a cosmic level, taking into account the universal viewpoint.

Illustration 1:
Dr. Eleanor Haspel-Portner

Twenty years ago, the Human Design system melded these four disciplines into one synergistic form that leveraged the truth contained in each, affording greater, more detailed information. Noble Sciences goes Beyond Human Design by integrating the sciences of Developmental Psychology and Clean Language into the Human Design System. This expanded information provides a complex personality tool that serves as a roadmap with depth and transformational power.

In 1987, Robert Allan Krakower, also known as Ra Uru Hu, received the mechanical structure of Human Design during an encounter with what he called "The Voice". He was given immensely detailed knowledge, especially of the body map, and was shown how to combine the four wisdom traditions into a single unifying, albeit incomplete system. Ra described himself as a mechanic and a messenger for the information he was given. He was provided with the structure, not the content or interpretation of the system he was shown.

Thus, one of Ra's goals when we met in 1996 was to validate the Human Design system. He approached my husband and me with that intention in 1999, and asked us to assume responsibility of investigating the Human Design system medically, psychologically, and scientifically. My statistical research, with the intention of validating Ra's hypotheses about "his" system, proved that it was a flawed system.

My work in Human Design, thus, led me to recognize that although it contained seeds of truth that resonated deeply within me, the data that it returned was not always aligned with the situation or clinical picture. For example, I noted key information that seemed to be missing from my childhood chart as well as that of my husband and children. I also searched for but couldn't find explanations in Human Design for some of the symptoms and issues clients presented.

Noble Sciences grew out of the synthesis presented in the Human Design system and my need, as a social scientist, to adjust it to get valid results in application to real individuals and their lives. Although I value the basic calculations and synthesis in Human Design, the method I developed in Noble Sciences goes beyond Ra's system by incorporating the social sciences and human development disciplines as well. These adjustments are my response to the shortcomings that I encountered and documented in the Human Design System when applying it to thousands of cases for statistical analysis.

In order to understand how and why Noble Sciences had to depart from Human Design and forge its own path, you need to hear some of my story.

My Story

My journey into synthesizing scientific and spiritual information began early in life. From my earliest childhood, I could see auras, and because of this, I was an avid student of esoteric traditions, practicing yoga spontaneously for as long as I can remember.

However, I was also drawn to the world of science and after receiving my undergraduate degree from Brooklyn College, I went on to get my PhD from the University of Chicago in the Department of Comparative Human Development. This was one of the first interdisciplinary social sciences departments in the United States, and afforded me the opportunity to study with some of the foremost psychologists, anthropologists, sociologists, and biologists in the academic world. I had the privilege of studying with people like Lawrence Kohlberg, Robert LeVine, William Henry, David Wylie, Bruno Bettelheim, Bernice Neugarten, Thomas Gordon, Erika Fromm, and Elizabeth Kubler-Ross, to name just some of the most highly-esteemed names in their fields.

Because I recognized the importance of theoretical and structural knowledge based on documentable research, I temporarily put some of my esoteric interests aside, and focused on my studies. While at the University of Chicago, I pursued clinical studies in Freudian psychoanalysis and Jungian analysis, transpersonal psychology, and religion. After my formal academic education, I studied meditation, astrology, metaphysics, and alternative medicine, as well. I was always drawn to open-minded disciplines grounded in scientific foundations, turning to them for answers to some of my questions about life transformations.

One of my life-long goals was to find a way to bridge the gap between science and esoteric awareness to make the world's wisdom traditions accessible. I wanted to find answers to questions about human life and nature that would explain the difference between those who lived happy integrated lives and those whose lives continued to be filled with struggles and challenges.

I always felt enlivened when I recognized the deep wisdom in others and acknowledged it in daily living. My driving motivation remained to find ways to facilitate change in people's lives without undermining their self-esteem or core being—something that can happen when people give away their power to esoteric or traditional teachers, courses of study, therapists, or educational systems. Within my work, I always served as a coach, partnering with my clients in their own evolutionary process.

In 1978 during a meditation, I was guided to travel to India. I had already listened to the "voice" of my soul and had successfully moved my family and clinical practice to California, so when I was guided to travel to India, I planned the trip. On that visit, and on the day predicted by my great astrology mentor, Katherine de Jersey, I met Marvin, the man who became my husband. The magic of that moment changed the course of both our lives.

At that time, Marvin had taken an indefinite leave from his holistic medical practice in California, had sold his home (a mile from where I had moved a year earlier), and was undecided about the future course of his life. In fact, Marvin was seriously considering remaining in India and living a meditative life. However, our bond of love was instantaneously deep and still continues today; we both knew we had met our true soul mate.

Marvin and I remain happily together in our life and work. We both honor the art and science that brings consciousness to those who are open to living courageously, honestly, and with integrity. We recognize that consciousness requires courage to change the circumstances and direction of life. We have, ourselves at times, made hard choices to honor our own life-purpose and paths. We know first-hand how hard change in life can be and the kind of courage it requires.

Noble and the Twins: A Story of Love and Inspiration

As any animal lover will agree, our pets become family members and can have an incredible impact on our lives. The name Noble Sciences was inspired by my cat, Noble, and the powerful relationship I had with him. Without Noble and some of the other cats in my family, I may never have found the key information and elements that were missing from Human Design.

Illustration 2: Noble

The connection between my work now and the cats in my life began in 1966 when I felt a deep drive to have a Siamese cat. I had never been in a relationship with a cat before. My search for the "perfect" cat led me to adopt a twelve week old kitten whom I eventually named Noble. Noble changed the lives of everyone he met. He was a blue point Siamese, who had a very strong presence even at that early age.

The first night Noble was with me, I was afraid that he would jump on me while I slept. As I prepared for bed, I put him on the floor, but he continued to jump on the bed several times. Finally, I lay down hoping he would understand my message. Instead, he hopped up on the foot of the bed and looked directly at me. I waited, this time, not moving. He proceeded to jump directly toward my face. I froze in fear.

He landed with a front paw on either side of my head. I looked directly at him, and he at me. The message he was sending me was clear and precise. Through his action Noble said, "I know what I am doing. I will never hurt you." I said aloud, "I understand." This defining moment of deep trust continued throughout our 22-year relationship while he was in his body. Our relationship continues to this day. I still feel his presence around me all of the time. He and I always respected our promise of trust made in that moment.

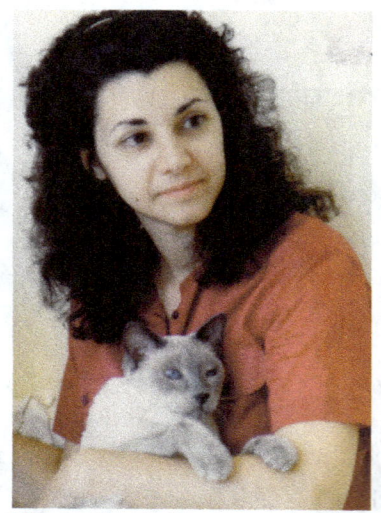

Illustration 3: Eleanor and Noble

During Noble's life, his awareness and healing powers were demonstrated daily. He protected me, my children, and my home, sometimes growling or "meowling" at any threat to us. Noble could intimidate people with his fierceness. Often, when I worked, Noble was at my side, telling me, through his actions, how to work with someone as well as how I needed to direct my energy. When I had clients in my office, Noble used to go to them, placing a paw on the area of their body where energy was blocked, and then he'd turn and look at me. This communication was witnessed many times by numerous people, especially when sessions were conducted with groups. Noble always knew exactly who to approach and where they needed help.

When my husband and I first returned from India, Marvin saw Noble on the dinner table. At first, he was surprised, stating that cats shouldn't be allowed there. Two days later, Marvin was in significant pain from a back problem. Noble found him lying on his stomach and leapt onto him, pressing his paw exactly against the place where

Marvin was feeling the most pain. Later, Marvin said that it felt as if there was a light and heat emanating from Noble's paws, healing his back instantaneously. After that, Marvin allowed Noble to go wherever he wanted.

Illustration 4: Joe and Ruzle

Just about everyone who met Noble got a cat. During Noble's last months, we were blessed to find a veterinarian who recognized his depth and powers. His veterinarian became and remains my dearest friend, a soul sister. After Noble's transition (July 23, 1987 @ 20:40 in Pacific Palisades, CA), this friend suggested that we consider the experience and joy that breeding a litter of kittens would bring. Such began a very powerful journey leading to the birth of Ruzle and Joe on January 18, 1990 @ 07:32 and 07:50 in Pacific Palisades, CA.

Ruzle and Joe were the first and second born in a litter of six kittens. All the kittens looked quite different from each other except for Ruzle and Joe, who were identical twins. During the kittens' births, we recorded the exact times they came into the world with a stopwatch so I could look at their astrological charts.

Illustration 5: Litter of Six Kittens

While I helped all six kittens and their very protective mother survive, in addition to weighing the kittens daily, I occasionally found a whisker, a strand of fur, or some slight variation that helped me tell

Ruzle apart from Joe. It wasn't easy. They had the same mannerisms, the same voice, and the same strong charismatic personality. In spite of this, there were a few differences.

When I began looking at Human Design Charts in 1996, I found it disturbing that all six kittens, born over about a two-hour period (07:32 to 09:43), seemed to have the same mammalian design charts.

Illustration 6:
Joe, Ruzle, Bear, Chester, Ted, and Rose

As is so often the case, my quest was emphasized when Joe became ill on April 14, 2000. He was diagnosed with congenital cardiomyopathy, a genetic condition. He died on August 31, 2000 @ 00:15 am in Pacific Palisades, CA. Ruzle continued to live a healthy life until December 31, 2007 @ 12:50 in Los Angeles, CA.

I knew there was missing information in the Human Design charts that could differentiate them and explain why Joe's life was so much shorter than Ruzle's. "The Voice" had not given Ra any information about Mammalian Design other than that when a Human sleeps he enters a mammalian-like matrix, a matrix that Ra also erroneously labeled the Dream Matrix. In Ra's system, the dream matrix and sleep matrix were synonymous. The problem with that is that they're actually two separate things.

Several distinct types of cycles occur when we sleep. During the deepest part of the cycle, the Sleep Cycle, all humans and mammals are without consciousness; they are in a physiological state of deep sleep where only the biological world is active. During this stage of sleep, a mammal is about as deeply unconscious as they can be without being in a coma. This is different from what happens when we go into Rapid Eye Movement (REM) sleep, otherwise known as the Dream Cycle. While dreaming, all 64 Gates are activated, creating a portal that allows us to receive communication from the unified integrated field of consciousness, or what Jung called "the collective unconscious". This phase of sleep enables us to transmit and use some of our unconscious information in waking life. It is during this phase of sleep that neurological communication between different states of consciousness occur.

It was clear to me that the difference between Joe's chart and Ruzle's chart had to be located in the Sleep Cycle, where their lives were affected on a biological level, rather than on a dream level. This fact led me, as a social scientist/psychologist, to research the differentiating calculations that would identify the personality characteristics of six distinctly different beings.

In 2000, I statistically analyzed the Human Design material looking for a marker that would account for Joe's illness and death. I found a very clear differentiating marker that told me what I needed to know to feel that the accuracy of the material could be described and be documented as scientific. When expanding the calculations to include the physiological or biological timing in the charts and in its corresponding matrix, I was able to determine that the position of the Moon in Joe and Ruzle's charts was different. I found that the 10-second difference in birth time in their biological charts shifted the chemistry in Joe and Ruzle's physical lives. I hypothesized that this difference accounted for the differences in their longevity.

**Illustration 7:
Noble**

I also wanted to identify and document the difference between the sleep cycle and the dream cycle in both humans and mammals.

With the help of Erik Memmert, who wrote the original program for the Human Design System, I found important differences in several key calculations that I now use in all Noble Sciences charts (refer to [Step 2: Understanding the 8 Charts](#) for better understanding).

Since discovering the information that Ruzle and Joe's births and deaths provided, the data that validated it has helped thousands of individuals to date. Furthermore, it has the promise of changing the lives of countless other humans and animals for the better in the future.

While I was looking for the missing markers for Joe and Ruzle, I began scientifically testing the Human Design System. In 1999 I used social science and astrological research tools to test the system on 30,000 cases. I had access to accurate birth data and research data on health and psychological conditions on these cases. I even compared the lives and charts of human twins, to validate what I'd already learned through Joe and Ruzle.

As I had found with the cats, my analysis proved that the Human Design System could be proven only in part because it focused on one calculation that addressed merely the mental world. Humans function in four worlds: the Mental, Spiritual, Emotional, and Physical. Their development and consciousness is not frozen at their time of birth. Thus, in any complete system, critical times in human development and how consciousness impacts a person must be considered and taken into account.

The results of my efforts to probe further into the animal charts led to greater understanding of human charts, and helped me expand the Human Design system into a more complete set of tools that provides the valid and reliable scientific findings of my research. I identified calculations and charts missing from the Human Design system. This was a labor of love. Thus, the new methodology I discovered and developed is called Noble Sciences in honor of my first cat, Noble.

As I continued working with the Noble Sciences' Charts and continued my own professional development, I recognized that correlations between the component disciplines synthesized in the Human Design graphic provide the roadmap to an accurate and powerful reading of the chart. The innovative field of Clean Language and Epistemological Metaphor as described by David Grove and Cei Davies Linn serves as a powerful way to clear individuals' predispositions and move them through their personality issues. The body graph provides the roadmap and "Clean" tools provide the process for clearing issues of the personality that connect individuals with their higher self without trauma or long term therapy.

Similarities between Noble Sciences and Human Design

- Noble Sciences moves beyond Human Design using it only as a structural component included in Noble Sciences information.
- Noble Sciences and Human Design use a similar graphic image or Mandala (a human figure with the chakras in the center of the astrological wheel surrounded by the I-Ching).
- Both synthesize information from the Tree of Life, the Hindu Chakras, Astrology, and the I-Ching.

Key Differences between Noble Sciences and Human Design

- While Noble Sciences and Human Design charts appear similar, the theoretical assumptions and hypotheses underlying them diverge significantly.
- The components and complexity of Noble Sciences differs significantly from Human Design. A Noble Sciences chart for

an individual or for a particular date or time is made up of a composite of more than eight different charts. A Human Design chart combines only two different charts. The additional calculations used in Noble Sciences Charts were discovered as a result of years of research; they are statistically verifiable, scientifically accurate, and chart the multidimensional layers of human and animal consciousness. For example, an individual may be a Manifestor in the Human Design chart but in Noble Sciences they may become a Generator in their spiritual chart. This would indicate that for them, right action depends on waiting for an inner voice of knowing before taking action. This has proven beneficial numerous times. In fact, my own chart provides documentation for this process.

- Noble Sciences documents five Types (or Five Ways of Being). Human Design hypothesizes and uses only four Types (or Ways of Being). This is important because the Ways of Being illustrate the manner in which you respond to the world around you and how you bring things into manifestation in your life. Keep in mind that individuals can manifest the things they don't want as well as the ones they do. Noble Sciences helps you make the right choices to consciously manifest your goals and bring them into your physical reality. (Note: For more information see The Five Types or The Five Ways of Being (Illustration 36.)
- Humans have four distinct Layers of functioning awareness: mental, emotional, spiritual, and physical. Human Design only takes into account the mental layer of functioning. The eight additional charts Noble Sciences uses address layers absent in the Human Design model. Each one deals with a specific layer of functioning. Two charts reveal the mental layer,

two the spiritual, two the emotional, and two the physical. In addition, there are two more composite charts that illustrate how an individual functions in an integrated way.
- Noble Sciences charts cover a six-month developmental period — three months before and three months after the actual date of birth to arrive at a comprehensive understanding of the individual or day in question. Human Design only covered a three-month pre-birth period of time and although they applied conception charts to their evaluations, the content provided never held when analyzed for accuracy. Noble Sciences' use of the six-month period yields greater detail and precision.
- In addition to the usual planets found in Astrological charts, Noble Sciences adds the planetary body Chiron into the map illustrating how a person functions. Noble Sciences' use of Chiron is an important detail. Although first noted in 1897, it was dismissed. Prior to its rediscovery in 1977, little astrological work with Chiron was done. After 1977, Chiron became important in astrology, especially in terms of relationships, marriage, and other gifts and talents in a chart. Chiron's use in charts and its application in astrology became invaluable. In my view, without it, understanding of one's chart is limited. Chiron enables the astrologer to determine an individual's higher life purpose and karmic gifts as well as how evolved the consciousness of the individual in question is. Human Design does not apply Chiron to its basic chart evaluations.
- Noble Sciences is the outcome of many years of research and development. While grounded in the wisdom of ancient traditions, its data is scientifically tested for accuracy on more

than 60,000 cases to date (2015). Despite claims of scientific testing, Human Design has not made adjustments to correct its misinformation.

- Noble Sciences incorporates developmental psychology principles into the timing of the charts and uses critical periods in infant development that are scientifically documented and congruent in the charts. Human Design doesn't use critical periods in their charts.

The calculations and charts used by Noble Sciences and Human Design set them apart as two systems. The two are also worlds apart in philosophical outlook. Human Design asserts that the birth moment locks a person into an inescapable way of living as defined by strategies of the Four Types (or Ways of Being). At its core, Human Design assumes that you have no choice in your unfolding process. It claims you most likely live your "not-self."

Noble Sciences believes information provided in its chart maps is useful precisely because the chart maps show predispositions about choices you make in your life. Noble Sciences Tools utilize cosmic energies as a guide in adjusting your thinking, feeling, and behavior to better align multidimensionally. It's helpful to know about tendencies based on your birth data so you can personalize and uncover your best choices uniquely suited for you within the context of the Universal energy flow on any given day and at all times.

Human Design stops short of providing practical guidance for each unfolding day, focusing instead solely on the birth moment as the defining pattern for all of your life.

Another key advance found in Noble Sciences is its use of whole brain tools such as mind mapping, physiologic exercises, meditative work, and EFT (Emotional Freedom Technique or tapping). These tools offer complex concepts underlying this work in a comprehensible format. These and many other useful tools will be covered later in the section entitled Applying Noble Sciences to Your Life. They are also available on the Noble Sciences website.

Noble Sciences' knowledge base is truly a sacred synthesis. My years of social science research, teaching, clinical, and coaching practices qualify me to remain at the cutting edge of science while exploring esoteric traditions. I apply them in modern practical applications that take psychological and coaching tools into a whole new arena of effective practice. With Noble Science as the only Multidimensional Personality system of its kind and me as its developer and innovator, you have access to knowledge never before revealed in its current depth of application and dimensionality.

Now you know a bit about the origins of Noble Sciences, its evolutionary process, and how it integrates ancient wisdom with scientific research. The next section explores the key components of Noble Sciences.

Applying Noble Sciences Sacred Synthesis to Your Life

Unpacking the Personal Toolbox

To help you integrate information provided by the Noble Sciences System, there are several important Tools to assist in your personal interpretation of the cosmic climate.

The Mandala

The Mandala, or chart, illustrates your active and inactive energy centers. It is more than a zodiac wheel: it is a map of the way cosmic energy flows at a specific time, based on the time used to calculate the positions of the planets at that moment based on where you are on earth. The Center of the Mandala graphically represents nine Centers of influence, common to all humans. A Hexagram corresponds to an astrological position of a planet at a moment in time. Once it is calculated and translated from an astrological position into a Hexagram and Line of the I-Ching it is put into a body map or graph.

When one Hexagram of an energy Center connects with a Hexagram of another energy Center, the connecting pathway is called a Channel. The connecting lines between the Centers clearly show the potential pathways or Channels through which energy may flow when a planet activates I-Ching Hexagrams; in this way, they describe its influence. One of the 64 I-Ching Hexagrams is associated with each of the Gates (energetic openings into a Center). (These are the numbers you see around the edges of each energy center in the close up diagram).

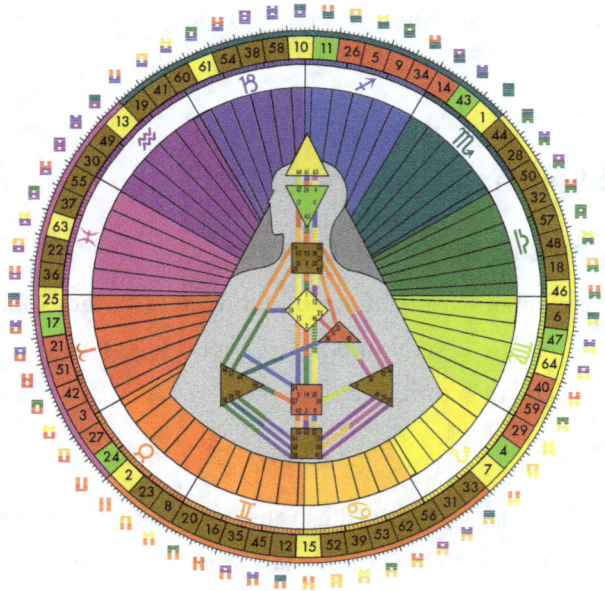

Illustration 8: Mandala of Synthesis

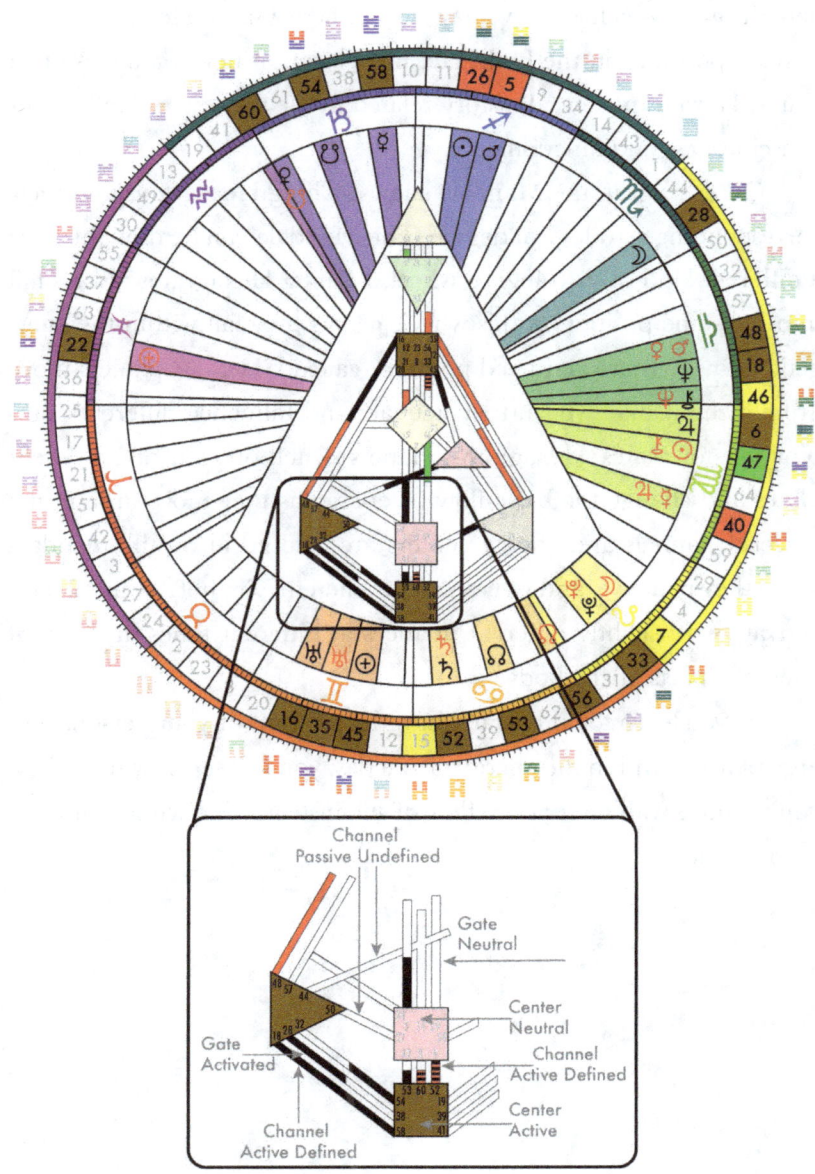

Illustration 9: Basic Chart Mandala with Definitions

There are 64 total Gates/Hexagrams and 36 connected Channels through which energy may flow. These Gates/Hexagrams are in fixed positions in the body, i.e., their location never changes. The Gates/Hexagrams reveal a story related to the way energy influences consciousness and functioning.

Outer rings in the Mandala show I-Ching Hexagrams and their corresponding astrological degree. The Mandala can be thought of as a different kind of astrological wheel. I-Ching Hexagrams always fall into the same position in this wheel, just as they fall within the energetic body Centers in a fixed position, called Gates. As planets move around the zodiac wheel, they activate and influence different combinations of Gates/Hexagrams. Planets switch on (activate) different Gates and change the basic flow of energy as they move around the wheel. Channels and energy centers are colored in on the mandala when they're active, and remain white when they're not. Thus, planets activate with slightly different nuances at different times in different energetic parts of the body.

As you learn to understand how the patterns of your personal energy play out within the energy of the day, you will see how they allow you to move with the natural flow of life instead of moving against its natural tide.

Ancient Roots: A Foundation of Wisdom

At the heart of Noble Sciences stand four of the world's most revered wisdom traditions: the Kabala's Tree of Life, the Hindu Chakra system, the ancient science of Astrology, and the Chinese wisdom system known as the I-Ching. These visionary mystical traditions have provided information and guidance to millions of people for thousands of years. To appreciate Noble Sciences' synthesis of these four esoteric systems, some basic understanding of them is necessary.

The Tree of Life

At the core of Kabalistic teachings is a diagram called The Tree of Life. It symbolically represents a process that describes the way Kabalists believe the universe came into being. It represents aspects of the Self and the journey back to oneness. Interestingly, the Kabalist view of the evolution or emanation of the universe is in many ways quite similar to explanations given by modern scientists. Kabalists believe that the universe began from a singular source, a dot in the center of nothingness that contained infinite energy.

It is thought that this energy went through a cooling stage as it expanded into space and time. The combination of the hot energy of the Big Bang and the cold energy of space created the material of the universe and formed the first atoms and galaxies. The universe operates in unfolding sequences that can be traced through certain pathways that form regular patterns.

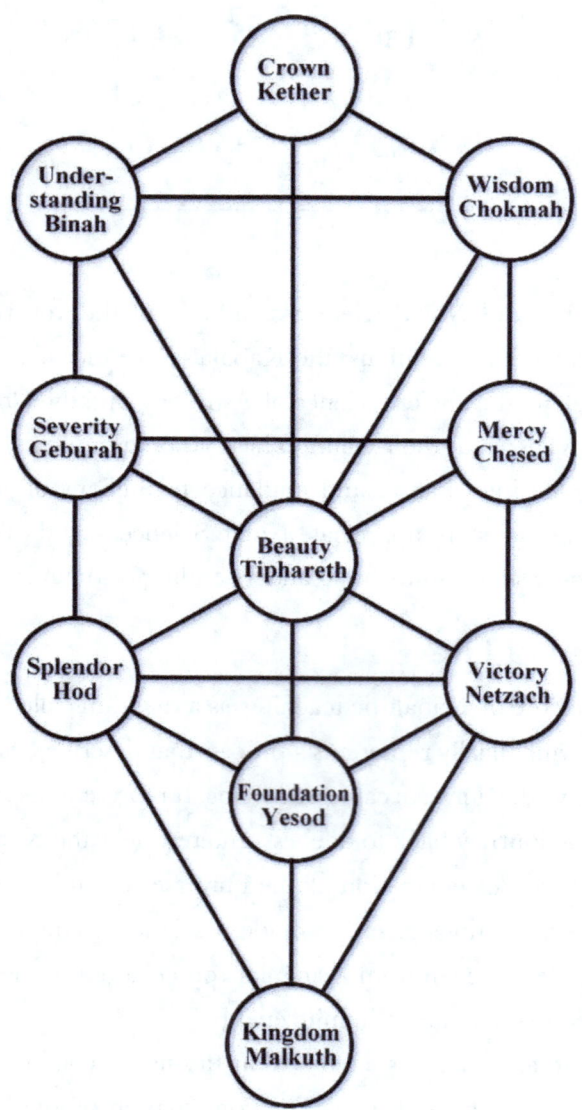

Illustration 10: Tree of Life

The Tree of Life usually shows ten energy centers connected by 22 interconnecting pathways. Each energy center carries its own sphere of influence within each of four auric bodies. Each energy center symbolically represents areas of universal and human consciousness that function in their own unique way in each of the four worlds.

The 32 Paths of Intelligence outlined in the Tree of Life relate to the 384 Lines of the I-Ching within the 360° astrology wheel. Thus it directly links together the components of the Noble Sciences synthesis. Understanding the relationship between the Paths of Intelligence on the Tree of Life and their integration provides a useful tool for understanding where and how life's issues affect behavior.

It has long been hypothesized that the paths of intelligence in the tree of life unfold in accord with the process of individuation, self-actualization, or psychological development as it unfolds in life. As an individual becomes increasingly aware and conscious, they evolve along a predetermined pattern that follows the paths of intelligence and result in increasing balance and harmony in that person's life. This sequencing is similar to the balance described in Taoist traditions and are in accord with the traditions of the I-Ching.

Hindu Chakras

Chakra Energy Centers are an ancient Eastern construct and tradition now finding widening acceptance in Western culture. These Energy Centers describe an organization of interlocking energy flow in the unfolding consciousness of all human beings, no matter where they reside on the planet or the culture from which they originate.

The word Chakra derives from Sanskrit; it means both a blossoming flower and a revolving wheel. In the West, chakras, or foci of energy in the body, are considered interior stars and are associated with precious alchemical metals. These energy hubs vitalize the physical body. They are associated with interactions on all layers of functioning: mental, emotional, spiritual and physical.

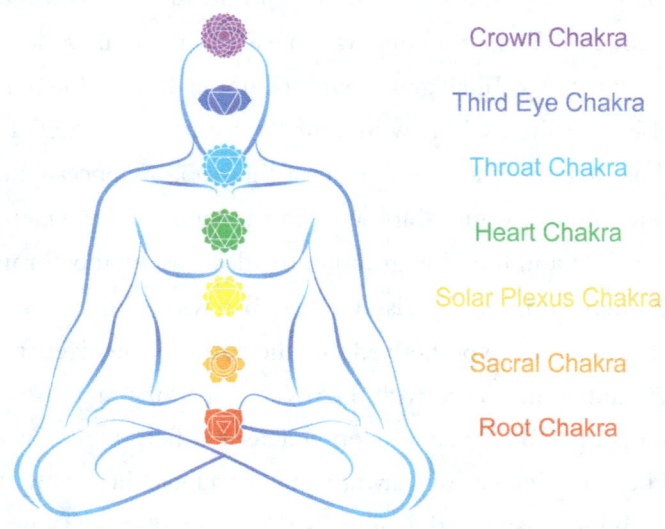

Illustration 11: Chakras

According to the Hindu Chakra system, within the human body are seven main spinning energy centers aligned in an ascending column from the base of the spine to the top of the head. The unfolding energy, called Kundalini, spirals up the spine in a rotating pattern.

Each chakra is associated with a certain color and sound frequency, and has specific functions, aspects of consciousness, and other distinguishing characteristics. Each rotating hub is the loci of life energy (chi, or prana) that flows through the energy Channels of the body. These energy vortices are involved in determining an individual's health and well-being. When all are functioning perfectly, the body and experiences of the individual also function with sound precision. When the chakras are dysfunctional, one can suffer illness, disease, or other misalignments in their material and day-to-day reality.

In working with the Noble Sciences' Charts, the Chakras or energy centers in the body graphic determine the kinds of issues and patterns a person might show as well as the world: mental, spiritual, emotional, or physical, impacted by the cosmic forces acting upon them in that world. Knowing the way energy varies within the body at different times makes navigating life circumstances easier.

Astrology

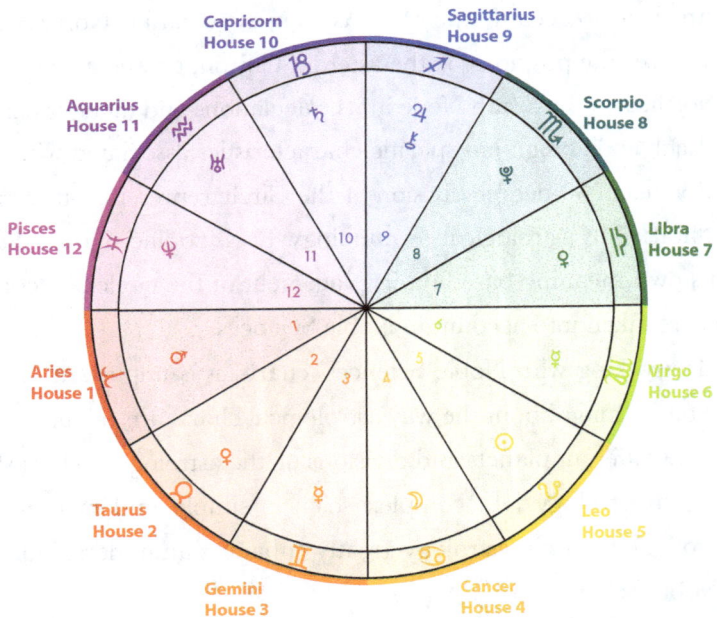

Illustration 12: Color Coded Astrological Wheel with Signs, Planets, and Houses
http://beyondhumandesign.com/go-beyond-human-design/ancient-wisdom/multidimensional-synthesis/

Dating back to Babylonian times, humans observed that planetary positions in the zodiac seemed to relate to human events. Through careful observation they noted recurrent planetary positions and developed certain hypotheses over time about individual characteristics associated with the planetary positions and relationships present at birth along with other key times of interest. These hypotheses coalesced into various forms of astrology that survive into current times.

Astrological charts and interpretations generally show 12 zodiac signs and 10 planetary positions.

Noble Sciences uses all planetary positions: the Sun, Moon, Mercury, Venus, Mars, Jupiter, Saturn, Neptune, Uranus, and Pluto as determining factors in an individual's character, development, and events that are likely to occur in their life. As mentioned earlier, Noble Sciences even uses the positions of the asteroid Chiron, as well as the Earth, the North, and the South Node in its calculations and in all its charts.

Each zodiac sign has specific characteristics associated with it. In addition, each 30-degree division of the Circle represents one sign, or one "house" in astrological terminology. Each zodiac sign, or house, has its own meaning based on its placement in the circle. All components are taken into account in Noble Sciences.

In working with Noble Sciences' charts, it is important to learn some basic things about the way astrological charts are calculated and what the different planets and divisions of the astrological chart show. Although astrology is a complex science, simple understanding of some of the basics in astrology greatly enhances an understanding of the Noble Sciences work.

The I-Ching

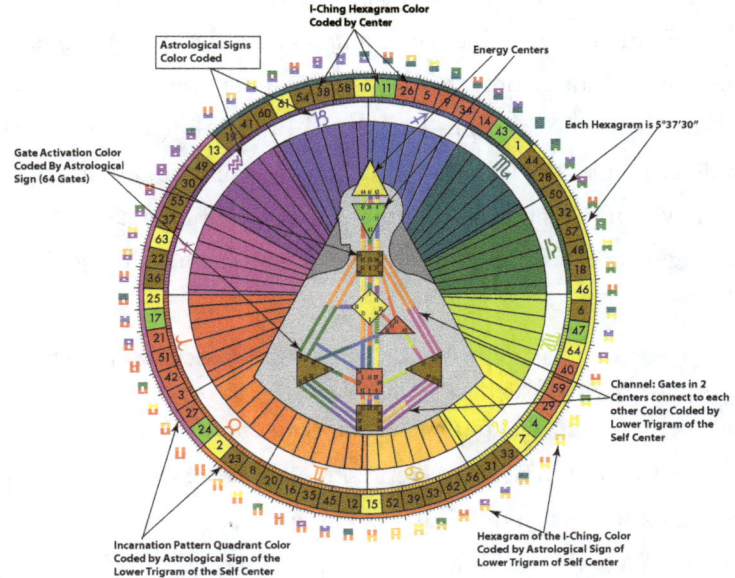

Illustration 13: Annotated 64-Gate Mandala
http://beyondhumandesign.com/go-beyond-human-design/
ancient-wisdom/multidimensional-synthesis/

The I-Ching, based on an ancient classic Chinese text, uses a system of images formed by stacking a series of six lines one upon

Illustration 14: Yin/Yang Hexagram

another. These six lines identify an order to seemingly chance events. Thousands of years ago, rulers in ancient China would consult the I-Ching for counsel, much like ancient Greeks consulted the Oracle at Delphi. Around the third century, a young scholar by the name of Wang Pi realized that the real value in the I-Ching would be found in using it as a tool for self-discovery. By developing a little personal understanding of the imagery and words used in the I-Ching, it is possible to advance your inner understanding of both the world around you and your personal response to it.

This ancient system forms a core of Chinese cultural beliefs. The philosophy of the I-Ching centers on the view of that life operates in a dynamic interplay of balancing energetic opposites. Yang energy, represented by an unbroken line, is masculine energy, described by the term Heaven. Yin, represented by a broken line, is feminine energy, described by the term Earth. These two energies are in constant motion, weaving in and out of our lives. The I-Ching teaches that the unfolding of events in any situation or process carries the inevitability of change.

Yang Energy	Unbroken line	——— Heaven	Masculine / Initiating
Yin Energy	Broken line	— — Earth	Feminine / Receptive

Illustration 15: Yin/Yang Chart

The I-Ching images are represented by 64 hexagrams, or patterns, of six lines each (384 Lines). These hexagrams each convey an energetic image that communicates wisdom passed through the ages. Each of the 64 hexagrams is made up of two trigrams. Each trigram is composed of one set of 3 lines that represent the balance of Yang, or masculine, active energy, and the Yin, or feminine, receptive energy. The meaning of each hexagram is determined by the pattern and arrangement of Yin and Yang energy in each symbolic image. The Yin (receptive/feminine) or Yang (active/masculine) balance of energy in any given trigram is based on the number of Yang lines that appear and the number of Yin lines that appear.

The chart below shows the eight trigrams that can be arranged in eight different combinations to make up the total of 64 hexagrams. Each of the eight trigrams of the I-Ching represents a visual image. Noble Sciences color-codes these I-Ching trigrams and maps each to its associated Astrological Sign, in accord with its primary sign.

Hexagram	Trigram & Sign Name		Astrological Sign	Keynotes
12	Heaven	♏	♊	Hexagram 12: Standstill: Caution (Mutation)
	Earth	♉		
11	Earth	♉	♐	Hexagram 11: Peace: Ideas (Acuity)
	Heaven	♏		

Illustration 16: Hexagram of Heaven-Earth Example

The Mandala of Synthesis

The chart below shows how combining the two Trigrams of Earth and Heaven results in two different Hexagrams depending on how they are combined.

I-Ching Trigram	Trigram Name	Astrological Glyph Image	Astrological Sign
☷	Earth	♉	Taurus
☶	Mountain	♋	Cancer
☵	Water	♌	Leo
☴	Wind	♎	Libra
☳	Thunder	♈	Aries
☲	Fire	♒	Aquarius
☱	Lake	♑	Capricorn
☰	Heaven	♏	Scorpio

Illustration 17: Eight Trigrams Astrological Sign and Name

In Noble Sciences, information calculated based on the time of your birth is used to create energetic body maps of the mental, spiritual, emotional and physical layers of your consciousness. The I-Ching Hexagrams, enhanced with Astrological information, form the core of this energetic map in a given state of consciousness. These maps, or personal Mandalas, show the pathways of energy through your consciousness.

Noble Sciences Charts Explained

In order for you to understand how this material works, you'll need a breakdown of some of the key terms in relation to the mandalas (or charts) used for interpretation in Noble Sciences. These explanations and definitions are important because they show the connection between science and esoteric theories, provide additional information about each of the eight charts I use in determining your dominant Way of Being, and show how to help you function best through the various layers of reality. We'll be discussing the eight charts in further detail later on, but the information they provide comes from the way the solar degrees and minutes and the lunar degrees and minutes are uniquely expressed for you.

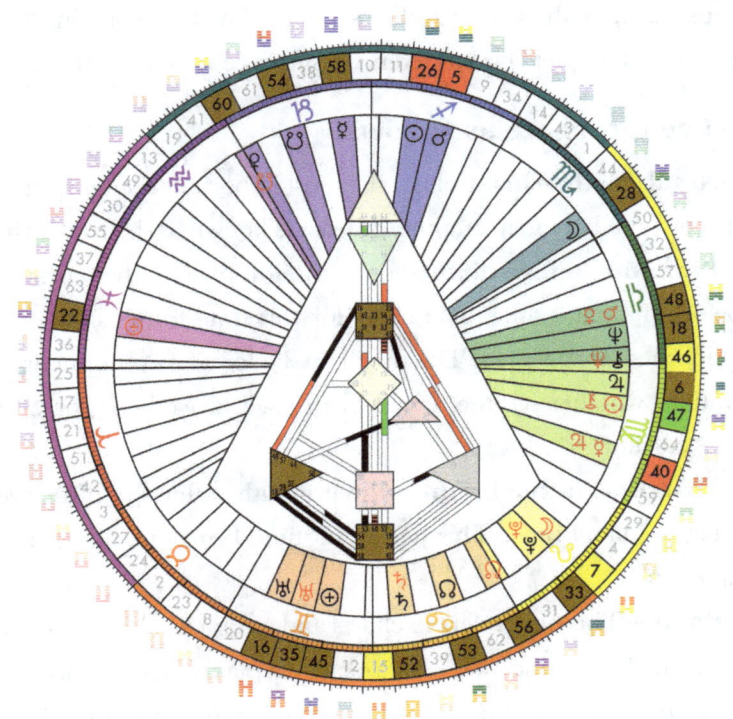

Illustration 18: Chart Mandala

The Importance of Birth Time

As you read, you will learn why it is important to look at the entire six-month developmental conditioning field to get a complete picture of energies at work at any given moment — current or otherwise. These energies are particularly important during the time surrounding the birth moment. The times for the charts and maps used in Noble Sciences have been chosen very carefully and scientifically. They correspond to key critical developmental moments in your birth journey and early development.

An astrological chart is like a clock that runs counter clockwise. Each degree contains 60 minutes, and within each minute is 60 seconds. Each of the twelve zodiac slices of the chart contains 30 degrees, so the complete circumference contains a total of 360 degrees.

The key terms are as follows:

Natal (birth) – In astrology, the most important event under consideration is called the natal time, otherwise known as the **birth time**.

Prenatal (before birth) – three months prior to the birth event, or before the natal time. The chart for this time frame shows when consciousness is activated. The time may be for any event under consideration, whether it's the actual birth itself, a wedding, or even a business deal.

Postnatal (after birth) – three months after the birth event, or after the natal time. This period of time brings manifestation to completion.

Solar – This time refers to charts and maps based on the degrees of the sun. Degrees indicate where exactly a planet is positioned in the 360 degree map of the heavens at a given moment in time. Because the astrological wheel is specific in terms of meaning in planetary degree such positioning reveals nuances about your personality, as well as events that might happen in your life. The solar position is the largest measurement, in terms of degrees considered from a planetary body, in the field of the developmental conditioning. Because of this, it has the biggest impact on every event and experience in our lives. The sun in your chart corresponds to the movement of the sun in the sky and affects you in the real world—revealing your *mental reality world*, how you interact in the world and function in relationship with your environment. There are several charts and maps based on calculations dependent on the degrees of the Sun.

Chart 1: 88 solar degrees prior to birth (about three months) – corresponds to your conscious waking reality chart. This is the point when a fetus might be viable outside the womb—or when the energy that formed the baby is capable of becoming manifest. Note that each solar degree is roughly equivalent to one day. It is also the time when consciousness becomes imprinted with information that registers in the developing cortex of the baby.

Chart 8: 88 solar degrees after birth (about three months) – corresponds to the chart that reveals your mental interaction with the world. This is the point where social scientists have observed that a baby possesses volitional cognition capacity. In other words, the baby is able to make a decision and act on it. For example, a baby might see a toy it wants and reach for it with intention. In life, this three month period is after the birth moment and allows any unconscious action to be corrected and adjusted according to the feedback within the three month period. It is essentially a learning and adjustment time for what will manifest fully.

Lunar – time based on the movement of the moon (its degrees of movement in the zodiac each day). This shows how your **spiritual or dream life** functions. One week after birth, the soul is anchored securely in the body (or the body undergoes "ensoulment"). Ritual ceremonies such as the bris in the Jewish tradition, and various other practices in numerous cultures mark the importance of the early period after birth.

Chart 2: 88 lunar degrees prior to birth (about one week) – corresponds to the your spiritual and dream world chart. This is the time when the fetus moves into position for birth or "drops." Its spirit seems to be preparing for birth and becomes impressionable and imprintable at this time. In practical life, it is a time when your unconscious intention gains momentum and is just beneath awareness. It is also a chart of the unifying field of creative intelligence and consciousness.

Chart 7: 88 lunar degrees after birth (about one week) – corresponds to the chart that illustrates how your spirit expresses itself. This is the period of time when esoteric teachings say the soul is anchored in the body and the also marks a moment that's valued in certain ceremonies within some religious traditions. Often it is an important waiting time for being certain that what you are thinking about doing is congruent with your deepest intentions and aligns with your spiritual life purpose.

Solar Minute – is a technical astrological term describing a short time (88 minutes of zodiac movement) before or after the natal time based on a shift in the Sun's position. It shows the timing of your emotional cycles and how you relate to bringing your consciousness into an effortless "angelic" state, or in other words, constantly being in the flow with the Universe. In this state, things fall into place on their own. The right job comes along, you meet the right people at the right time, and even find perfect parking just as you need it.

Chart 3: 88 solar minutes prior to birth (about two days) – corresponds to the chart that shows the true emotional reality you experience within in spite of how the world sees you. This is the point when labor activates with an intensification of Braxton-Hicks contractions. In other words, this is the moment at which the baby begins to make its way from the mother's body into the physical world.

Chart 6: 88 solar minutes after birth (about two days) – corresponds to the chart that reveals how you interact in close emotional relationships with others. This is the point when the mother's milk comes in and the baby's body is more fully engaged, connecting, and bonding in its attachment and relationship as a separate being.

Lunar Minute – is a technical astrological term that shows a time before or after an event based on a small shift in the Moon's position in the zodiac. It illustrates the way you function hormonally and are triggered physiologically, relating to the Sleep Design chart. This layer is usually beneath your awareness, functioning in the biological world.

Chart 4: 88 lunar minutes prior to birth (about 1.5 hours) – corresponds to the chart that reveals your unique physical and hormonal timer. This is the onset of transition—the final portion of the birth process.

Chart 5: 88 lunar minutes after birth (about 1.5 hours) – also corresponds to the chart relating to your physical and hormonal timer. This is the point for activation of the baby's body independently of the mother in its own hormonal waking/sleep cycle rhythm.

The Importance of Birth Location

In astrology the exact birth time, date, and place (latitude and longitude) of a birth affects the accuracy of the information charted. The reason for this precision is based on the fact that the earth's rotational speed is time sensitive in measuring planetary positions in the zodiac. Even seconds can make a difference in the information determined in the charts.

In research there is a saying, "garbage in, garbage out." This means that the data analyzed is only as good as the data collected and entered for analysis. As I learned when I analyzed data for my doctoral dissertation, the method for coding, analyzing, and interpreting data can vary with each person conducting the research. Good social scientific research documentation requires the utmost precision in terms of what is collected for analysis, how it is organized and coded, how the data is computed, and how the results are interpreted. Astrological data and chart analysis must, therefore, be accurate and properly interpreted to be of true value to those receiving the evaluation.

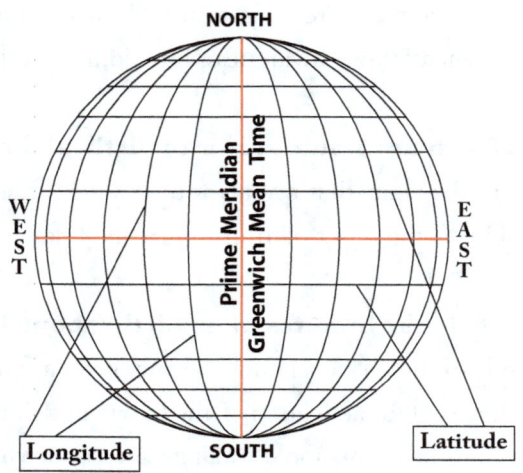

Illustration 19: Latitude-Longitude

The Earth is geographically divided into two imaginary sets of circles. One set of circles measures the distance north or south of the equator and is called latitude. The other set of imaginary circles divides the Earth from pole to pole into what is called meridians of longitude. All places with a given longitude have the same time zone, i.e., noon is the same no matter how far north or south you are from the equator. The lines are numbered for the meridians of longitude going east or west of Greenwich, England (a place arbitrarily set as the Prime Meridian) as the 0 point from which to measure time zones.

The equator is the place from which we move north or south to designate a location of latitude. By noting the latitude and longitude of a place exactly, the precise location of birth can be noted. In addition, the exact angles or perspective of the planets and their alignment in relationship to the zodiac and to each other can be charted specific to a precise moment in time.

# Unlocking Noble Sciences Sacred Synthesis

Noble Sciences is a multidimensional system of interlocking matrices that show how you function as an integrated being in a context of planetary influences. Because you operate in a cosmic program that transcends your existence, it is possible to see and chart the flow of energy over the course of days, weeks, and years to "read" the forces acting and interacting upon you and on the collective level as well. Not only can you get to know yourself better using the system, but you can gain deep insights about Universal happenings on a daily basis. As you will see, this goes far beyond any astrology reading based on the zodiac.

Personal and Universal:
The Field of the Conditioning Climate

The detailed information provided by Noble Sciences can be applied to you based on your birth data or it can be looked at as a general "climate" for a given day as it affects everyone. It is both a personal system that identifies tendencies within you based on your birth moment and it is a universal system that identifies tendencies that apply to you and other people on a daily basis.

My research revealed that looking at the day in question is not enough to get the full picture. The data revealed that there is a six-month "conditioning field" both for individuals and for any moment in time that extends from three months prior to the moment in question through three months after. Consequently, the energy of any given day or birth moment actually encompasses a period of six-months.

Like a very slowly developing picture, you can't fully comprehend what you see until the process is complete. And like the photo, it is a composite of several layers of color working together to make a complete picture.

The energy climate during the time and place of your birth, which includes the three months before and after, influences you at your deepest level. These energies coalesce together into key aspects that shape you. When you are older, the climate of each day still affects you, but you experience it through filters set up within you during the initial energy imprint you received during the key six-month developmental conditioning period surrounding your birth. These key times of development correspond to the eight charts used in Noble Sciences.

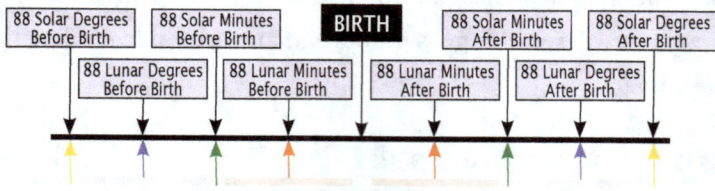

Illustration 20: Developmental Timing

Note regarding the chart above: Each of these segments of time relates to a layer of our reality. Solar Degrees affect the mental layer and Lunar Degrees affect the spiritual layer. Solar Minutes affect the Emotional (Angelic) layer, and Lunar Minutes affect the physical. These differences determine which world or matrix is used to determine how an individual functions in different expressions of their reality, as each world functions within its own timeframe.

The Developmental Conditioning Field gives rise to a multidimensional picture of a person or day. To fully understand this multidimensional picture, it is important to know the many factors or components that go into any single composite view. In this section, we will examine each of these key components:

1. The Four Layers of Functioning

Everyone functions through different senses—mental, emotional, spiritual, and physical layers. The Four Layers of Functioning are found in many disciplines and across many esoteric traditions, often referred to as subtle bodies or energy bodies.

2. The Eight Charts

Noble Sciences has pinpointed the times when each of the four layers of being are activated for any given birth/day and time. Each of these times produces a chart/map that, when looked at in composite, gives a detailed and clear picture of an individual person's make up, or of the cosmic energies of a particular day.

3. The Nine Energy Centers

The charts allow you to see which of the Nine Energy Centers are defined, or "turned on," over the six-month development period. You can also see connections between centers. The activations for the Nine Energy Centers and the ways in which energy connects these centers are critical components to understanding your inner process or the energy of a particular day.

4. The Five Types

Because information from charts and maps represent each of the four layers of functioning and shows how and when the Nine Energy Centers are activated, it is possible to determine which of five basic Types you (or a day) falls into. By looking at the flow of what is activated or deactivated, it is further possible to ascertain (decode) exactly the way you or the day functions.

Step 1: Understanding the Four Layers of Being

There are many layers of functioning, but those who only believe in the physical and personality expressions of being are limited and missing out on learning from the other realities they actually function in unconsciously.

As a human being, you function in four worlds. The **mental** world determines how you **think** about things in your life, your relationships to others, i.e., your daily reality. The **spiritual** world determines how and what you **perceive** as your purpose for being in the world as part of a greater whole. The **emotional** world determines how you **feel** about things that relate to you in all areas of your daily life. The **physical** world determines your **physical** health and how its chemistry sustains your life.

Each of these worlds operates within you all the time. Sometimes, I describe them as layers since that is how they are generally perceived. These layers, or worlds, correspond to the energy bodies or auric fields often identified by mystics and esoteric groups. We are a microcosmic reflection of the macrocosmic universe we live in, therefore, our energies interact in a dynamic fashion with the world around us. If we are energetic beings, then the thoughts and feelings we experience in our subtle bodies can eventually be translated to manifestation in our physical realities as these frequencies become denser. Regardless of the way you live your life and regardless of your cultural practices, these are common energetic layers we all experience. When these areas of your life are in balance, you're likely to feel a sense of alignment and happiness. When they're out of balance you are likely to feel a sense of disharmony or dissatisfaction.

These layers are activated during development three months before a birth event. This time marks the time when your consciousness becomes viable. It extends up to your time of birth. Once birth occurs, the environment and those in it affect the way you respond developmentally and how you are socialized in and to the world around you, especially, in the first three months after birth.

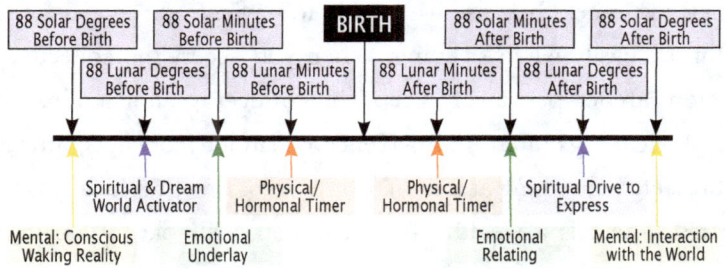

Illustration 21: Timing in Birth Chart

Patterns of development influence your ways of responding and perceiving. Your patterns become conscious when you build awareness. By knowing how you function in the four different worlds, and at different times, you gain awareness of your predisposed ways of perceiving. Awareness leads to the ability to change your predispositions and unconscious behavior.

The Waking World (Mental and Interactive)

Your mental or waking world may appear to be the most powerful of the four layers of functioning because it may seem to be the most "real." This world is where you function most of your waking day. It is indeed powerful because the way you perceive ordinary life happenings (reality) creates the world in which you live and interact. In fact, you may often operate as if it's the only dimension in which you live.

The Spiritual and Dream World (The Unifying Integrative Field of Consciousness)

The world of collective unconscious archetypes (symbols) activates when you sleep. While you are not using your mental or conscious mind, you still function in other ways. This world operates on the subconscious awareness and connects to your superconscious (higher) awareness because we're able to navigate through space and time more easily when sleeping. When you access the collective unconscious archetypes during sleep, you connect to all life. This is why archeological digs often find evidence of similar concepts, advances, symbols, etc., emerging at a similar time across geographically isolated cultures. There is a wealth of information available to you in your dream state and in your spiritual functioning if you know how to access it. This world connects you to your highest purpose and often gives you a sense of "destiny."

The Emotional World (Angelic)

In Noble Sciences, the emotional world is connected to the angelic realm. This is because when we experience higher forms of emotions such as joy or tranquility (supernal emotions), spiritual connection can occur and we can receive inspiration and insight that can be brought into conscious experience. As you can see, this state involves more than just what you normally think of as "emotions," this is the world that gives you meaning in your life. The manner in which you handle your emotional reactions to the world around you affects all the components of your life. Knowledge and mastery of your emotional functioning offers you the capacity to rise above the momentary reactivity of life and find pathways that connect you to your Higher Self. This is how you transcend choices made in fear, anger, and other feelings.

The Physical World (Biological)

You live in a physical body that operates from the moment of your conception. Your body is influenced by your experiences, but it too has influence over different aspects of your existence. Though you are always well aware that you have a body, you may sometimes disconnect from your body's intelligence and your cognitive mind. The Hormonal/Physical Regulator of your body determines your physiology, its chemistry, and timing. By understanding information in your physical world, you can better know how your body regulates and affects your health. You can learn where and how you are vulnerable and how to optimize your health.

Step 2: Understanding the Eight Charts

The multidimensional charts for any given date immediately show the progressive energy movement affecting that date. It reveals the unique way you experience it as an individual. Planetary movement is predictable and can be seen well in advance. Thus, it is possible to "look ahead" and also to "look back." By considering the full six-month developmental cycle of a birth moment, or of a given day, it is possible to get a clear picture in the "now moment" (or at the mid-point) and thus to anticipate and make choices that affect the way developmental process is completed.

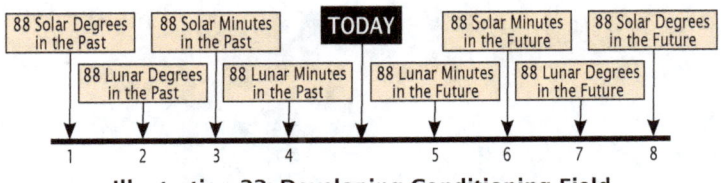

Illustration 22: Developing Conditioning Field

What often strikes me, beyond my ability to explain it, is the intelligence of the cosmic movement in seeming to coordinate the underlying harmonization of the planets. Planets move in regular patterns bringing awareness to the surface in ways that allow you to optimally balance mentally, spiritually, emotionally, and physically. At times, one layer of awareness is emphasized over another, with the emphasis on the inner process, while at other times, on outer activity.

The uniqueness of your own chemistry was influenced over a critical six-month period spanning the last trimester of pregnancy and the first three months of your life. During this time, your cognition (awareness) was activated in accord with your development in all four layers of being. To feel balanced and whole, you must come into alignment both as a unique person and as part of the greater whole.

Step 3: Understanding the Nine Body Centers

Noble Sciences charts document the activity in energy centers at key points in time (up to three months before and after birth) in order to see which energy centers come into play at key developmental moments. Each energy center represents certain aspects of the human experience, or characteristics of being human. You will notice that seven of the centers correspond to the seven main Hindu Chakras. In addition to these seven, there are two other key energy centers, the Splenic Center, which provides intuition about the health and/or well-being of the body, and the Self Center, which is the bridge between higher consciousness and personality consciousness.

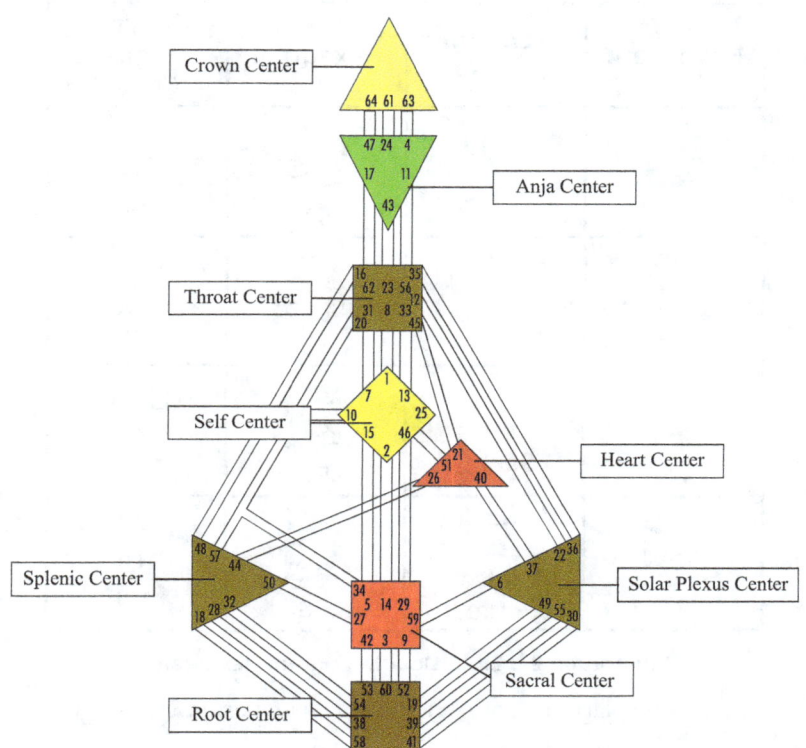

Illustration 23: Body Map with Centers Labeled

I-Ching and Hexagrams

When studying the Mandala Wheel with its hexagrams you might notice that eight of the 64 hexagrams in the I-Ching have both the same upper and lower trigrams. The eight double trigram hexagrams are strong energetic hexagrams and occur only in cardinal and fixed astrological signs. Notice also that there are no Double Trigram hexagrams in the Crown, Ajna, and Throat Centers. This is because Double Trigram Hexagrams represent energy that still needs to be mastered, however, the energy centers that are related to higher activity do not require this type of learning.

Hexagram	Fixed Signs Center	Hexagram	Cardinal Signs Center
1	Scorpio ♏ Self	51	Aries ♈ Heart
2	Taurus ♉ Self	52	Cancer ♋ Root
29	Leo ♌ Sacral	57	Libra ♎ Spleen
30	Aquarius ♒ Solar Plexus	58	Capricorn ♑ Root

Illustration 24: Eight Double Trigram Hexagrams

Noting the placement and energies of these hexagrams is significant because they emphasize the energy located there and drive it through the centers in which they occur when active in strong ways. Think of them like an exclamation point in the energy center, heightening their power when they are activated. In addition, the experience of these hexagrams is internally strong in the person in whom they are actively defined.

When looking at Noble Sciences Charts it is important to recognize that each world has its own unique energy expressions and requires different language to describe the way the energy is experienced when activated in each of the worlds. A human being lives in all four worlds. Thus, there are descriptions for all Gates in the conscious waking reality world, the world of conscious interaction as well as for the Gates or hexagrams in the world of spirit, world of emotions (angelic world), and world of biological form.

When two worlds meet, the Gates of each world is influenced by the other world's Gate and functions much as a portal functions between two worlds. In addition, some Gates occur in multiple worlds and thus can be active as a bridge bringing information into both worlds rather than just connecting one world to another. Imagine that when you go through a portal you move from one consciousness to another with the essence of one coming into the presence of the other. The experience of a portal is similar to the experience of yourself when you are with a close friend versus when you are with strangers. You are the same person but you operate differently.

When there is a bridge between worlds, both types of awareness function simultaneously combining their different energies and actively energizing their expression. Thus, it is similar to how you might feel if you are relating both to your partner and your children at the same time, facilitating their understanding of each other and translating their meaning and context to each other.

The centers can also be grouped into different types:

Awareness Centers carry conscious energy. The energy hubs in this category are involved in our thoughts, subconsciousness, and intuitive wisdom.

Awareness Centers (3)	Motor Centers (4)
Ajna Center Splenic Center Solar Plexus Center (also a Motor Center)	Solar Plexus Center (also an Awareness Center) Sacral Center Root Center Heart Center

Pressure (Fuel) Centers (2)	Other Centers (2)
Crown (Head) Center Root Center (also a Motor Center)	Throat Center Self Center

Illustration 25: Centers Chart

Pressure or Fuel Centers contain energy that comes and goes. Sometimes, the energy collects in a fuel center and eventually builds to the point where it needs an outlet. If a pressure center is unable to find an outlet, the energy can manifest as disease and/or the inability to create and bring things into physical existence.

Motor Centers behave like an ignition. They provide the spark that sends energy through the right Channels and hubs in order to bring things into manifestation.

Any center that can be a manifesting source is going to want expression. It fuels action and motivation and can be turned on to make something happen. If the solar plexus is acting as a manifesting source, it will be your emotional energy that's moving. If your sacral center is functioning in this manner, your instinct will be active, and if your heart wants to manifest something, it's your will that will seek expression.

Now we will look in detail at each of the nine centers.

The Nine Centers

When looking at a chart, it is important to observe which centers are colored in or defined across the six-month snapshot. Some centers may be defined three months prior to your birth but switch to being undefined after your birth. Noting these patterns of changing definitions allows for a deeper understanding of your personal make up or how you interface with the day you are seeking to analyze. When centers "turn on and off" they tell you which layer(s) of functioning is (are) affected: mental, spiritual, emotional, or physical. Your basic birth charts spanning the 6-month developmental conditioning fields serves as the predisposing filter through which you know your Self and how you naturally relate to others and to the world.

The Crown Center

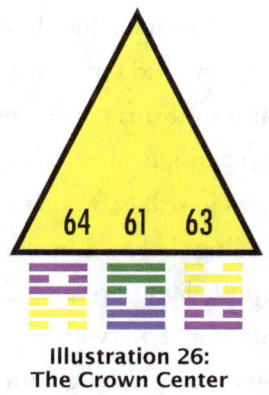

Illustration 26: The Crown Center

The Crown Center (Head) carries pressure to comprehend and make sense out of things, i.e. the fuel to conceptualize. It is where your mental activity begins. Energy activates your Crown Center in three ways: in a cycle, in a pulse, or in a focused way. These three forms of energy activation are instrumental in determining how energy movement is processed, perceived, and filtered by you.

At this location, energy works like light in the sense that sometimes it's a particle and sometimes it's a wave. It also behaves like electrical circuitry, as an AC/DC current. When energy is in cycle mode, you may start something, but never finish it. When energy acts a pulse, it switches off and on. You may have a brilliant idea one moment and then forget it the next. Energy that's functioning in focus mode is single-pointed. In this form, you can concentrate your thoughts with laser precision.

The **Root Center** is the other Pressure Center with similar ways of activating energy: As above, so below. When the Crown Center is colored in on your chart, this means it's in an activated state, and you will feel mental pressure to grasp and understand things in a certain way. A colored in Crown Center moves you to seek answers. When the Crown Center is not colored in, you'll feel inspiration and ideas from all different directions. Without a defined (or activated) Crown Center you are likely to live in the questions, i.e., ask more questions than you can answer.

The Ajna Center

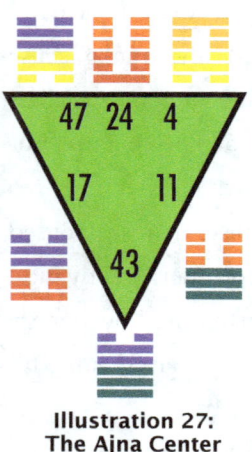

Illustration 27:
The Ajna Center

The Ajna Center is located just between and above the eyebrows and serves as a very important hub that transforms cognitive information into comprehension. The Ajna Center grounds your ideas in reality by assisting you in creating meaning. Located in the Ajna Center are the hypothalamus, pituitary, and other important autonomic nervous system structures. The Ajna Center, along with the Crown Center, functions to help you interpret information. You often make decisions based on this Center. The Ajna Center in its optimal functioning works in concert with the Sacral Center to interpret information from your "gut" feelings. It has the unique capacity to transform information from pure "gut" instinct into higher octave cognitive information—it provides deeper and nuanced meaning. It is this very crucial capacity that differentiates you, as a human, from other species. The Ajna Center is one of your three Awareness Centers (the others being the Splenic and the Solar Plexus Centers).

The Throat Center

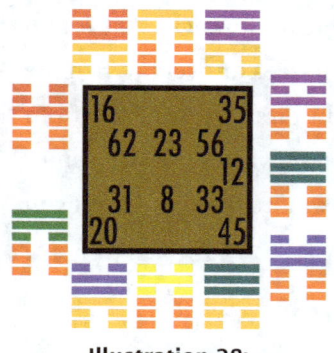

Illustration 28:
The Throat Center

The Throat Center, when colored in on the chart, facilitates self-expression either in words, through action, or both. The Throat Center is important in determining your way of being — the Five Types: Manifestor, Generator, Manifesting Generator, Projector, or Reflector

— as well as the way you communicate. The Five Types reveal the way you express yourself in life. The Throat Center is the most complex of all Centers. When the Throat Center is colored in and connects to any Motor Center directly or indirectly, it defines a Manifestor. When the Throat Center is colored in along with the **Sacral Center** and it connects to a Motor Center (defined above), it defines a Manifesting Generator, i.e., if you are defined this way, you can make things happen. When the Throat Center is uncolored, it indicates an openness to speak or act under the influence of the context of the situation. The Throat Center is essential in your capacity to communicate and to take action in the world. The Throat Center is a very complex and strong determiner of how you speak and act. (*Note: The Five Types or Ways of Being are discussed in more detail later.)

The Self Center

Illustration 29: The Self Center

The Self Center contains the core of your identity, your Higher Self. It is the magnetic force field that holds your transcendent soul and human being component together in time and space. The eight Gates in this Center set the roles and perspectives in the way you orient yourself in life. There are eight Gates in the Self Center, each having a different Lower Trigram of an I-Ching Hexagram. Only two Upper Trigrams of I-Ching Hexagrams are represented in the Self Center. The I-Ching Hexagrams fall primarily in the degrees of eight different Astrological Signs and give an indication of the

Self-Orientation or Roles and the way you express them in your life. Each lower trigram in this center has a double trigram Hexagram in your energetic body, meaning that the trigrams within the hexagrams are mirror images of themselves — they're repetitive. The Self Center when colored in shows that you have a sense of self-identity that is more defined than it is when the Self Center is uncolored.

Two double Trigram Hexagrams are in the Self Center, **Hexagram 1 (Heaven/Heaven) "The Creative"** is in the astrological sign of **Scorpio (♏)** and **Hexagram 2 (Earth/Earth) "The Receptive"** is in the astrological sign of **Taurus (♉)**. Lesson: This combination enables an individual to receive creative inspiration and make it manifest in their physical reality.

The Sacral Center

Illustration 30:
The Sacral Center

The Sacral Center is a primary source of energy within you, empowering your life itself. The Sacral Center generates the genetic imperative for reproduction, intimacy, and nurturing of the species. Within this center resides the Life Force energy also known in esoteric circles as Kundalini energy. It drives you toward survival. It also interacts with your cognitive understanding and gives you the intelligence to interpret your instincts. **The Sacral Center** coordinates with the **Ajna Center** in optimal functioning so information transforms from pure "gut" instinct into higher octave cognitive information. It is this capacity of integration between the Sacral and the Ajna Centers that makes humans unique among those on Earth. It is from the Sacral Center that you feel "gut" feelings and respond to things instinctively. It responds to internal as well as to external stimuli. When colored in on your chart, it is easier to know your "gut" responses. When not colored in, it may take more time to get clarity on your unique instinctive responses.

The double Trigram **Hexagram 29, Water/Water "The Abysmal"** is in the astrological sign of **Leo (♌)**. The lesson in this hexagram's configuration is symbolic of an individual plunging into an abyss of danger in order to learn. Once the lesson has been fully absorbed, they will rise to the surface into the light to succeed with renewed purpose.

The Root Center

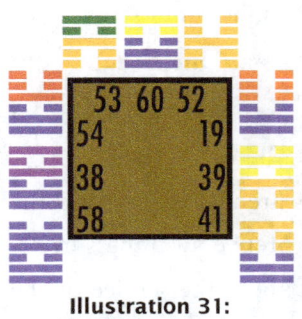

Illustration 31:
The Root Center

The Root Center is both a Motor and a Fuel Center. It processes the movement of energy through you. Its energies set in motion the way your body feels pressure. The Root Center activates energy. Like the **Crown Center** three kinds of energy originate in the **Root Center** determining the way the energy moves and is processed and perceived: cyclic, pulsating, or focused energy are possible. Again, "as above, so below." When this center is colored in on your chart you may feel internal motivational pressure that causes you stress if not acted upon. When uncolored you may feel unmotivated unless something external triggers you. The Root Center adrenalizes energy essential for all areas of activity in your life, grounding you in your foundation.

Two double Trigram Hexagrams are present in the Root Center: **Mountain/Mountain Hexagram 52 "Keeping Still"** is in the astrological sign of **Cancer (♋)** and **Hexagram 58 Lake/Lake "The Joyous"** is in the astrological sign of **Capricorn (♑)**. The lesson in this combination is symbolic of an individual's ability to find true joy through tapping into the stillness within. Patience brings bliss.

The Splenic Center

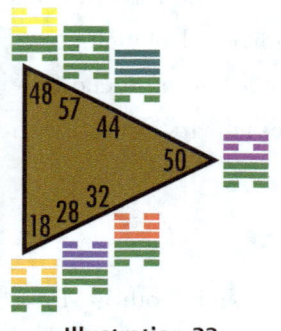

Illustration 32:
The Splenic Center

The Splenic Center functions when your body alerts you to conditions both internal and external as they affect your health. It operates as an awareness center. The Splenic Center joins with other key centers that work together to create the feeling/instinct/mind and body/instinct/mind axes of awareness; the centers are the **Sacral Center**, the Mind (the **Ajna** and **The Crown Center**), the **Splenic Center**, and the **Solar Plexus Center**.

The Axes of Awareness are the three awareness centers and the Sacral Center operating on two axes of interpretive awareness in determining if feelings have a basis in fact and are in the service of the higher self or not. The Axes of Awareness and understanding the way it operates is extremely important in the evolution of consciousness and in living in a balanced and harmonious way.

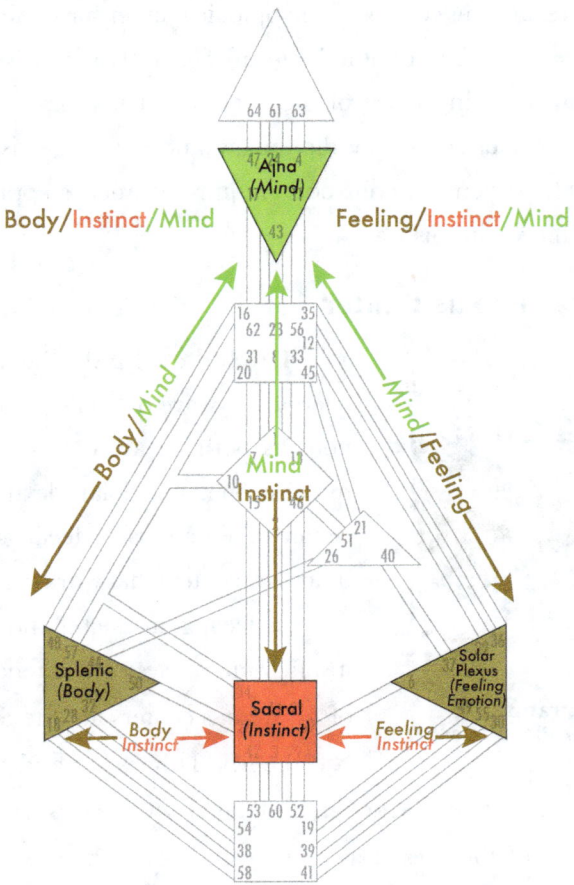

Illustration 33: The Axes of Awareness, Three Awareness Centers

    The Splenic Center, the body/instinct/mind axis, attunes your awareness to what affects health as it is modulated by instinct and cognition. When colored in on your chart, health is optimized and it's easier to feel spontaneous because your body instinct gives more reliable information in each moment than when your Splenic Center is not colored in. When you have an uncolored Splenic Center you are likely to be highly sensitive and to pick up in your body what others feel. This "open" Splenic Center often shows high empathic abilities.

All Hexagrams/Gates of the Splenic Center have the same lower trigram **(Wind)**. The double Trigram **Hexagram 57 Wind/Wind "The Gentle"** is in the astrological sign of **Libra (♎)**. Lesson: This combination illustrates how diplomacy and gentle persistence bring success. The systems in your body require balance, so optimal health can be achieved in this way.

The Solar Plexus Center

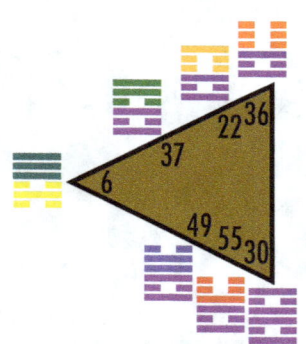

Illustration 34: The Solar Plexus Center

The Solar Plexus Center activates energy in a wave or cyclic pattern. It functions as both an awareness center and a motor center. The Solar Plexus Center, the center of emotions or feelings, joins with other key centers that work together to create the feeling/instinct/mind and body/instinct/mind axes of awareness; the axes of awareness centers are the **Sacral Center**, the Mind (the **Ajna** and **The Crown Centers**), the **Splenic Center**, and the **Solar Plexus Center**.

As part of the axes of awareness, the **Solar Plexus Center** controls the feeling/instinct/mind axis. It attunes your awareness to what affects your feelings, i.e., emotions as modulated by instinct and cognition. At this energy center, emotions behave like a wave in the sense that they start, surface, intensify, and then fade. When integrated, your emotions are balanced and their expression is optimized.

The **Solar Plexus Center** has impact on everyone, regardless of whether it is colored in or not because of the nature of human consciousness and of human nature. At this time in evolution, humans experience emotions and then respond to them. Someone with a colored in or actively defined **Solar Plexus Center** may experience emotional cycles and the intensity of feelings physiologically. They may also be more difficult to modulate cognitively than when the Solar Plexus is not colored in. Individuals with a **Solar Plexus Center** that is open, or not colored in, generally experience what others feel emotionally; they are emotionally empathic and may have some difficulty knowing whose feelings they are feeling. In other words, someone with an active **Solar Plexus Center** feels shifts in their emotions based on their inner physiological responses and desires, whereas someone with an open **Solar Plexus Center** feels emotions that may arise from their compassionate taking in of what those around them feel. All centers have challenges whether defined, i.e., colored in or undefined, i.e., not colored in.

All Gates in the **Solar Plexus Center** except for Hexagram/Gate 6 have the same Lower Trigram with the double Trigram Hexagram in the **Solar Plexus Center** being **Hexagram 30 (Fire/Fire) "Clinging Fire"** in the astrological sign of **Aquarius (♒)**. Lesson: This configuration shows how emotional awareness can make things better by shining light on the darkness.

The Heart Center

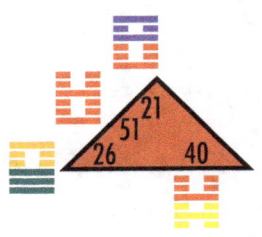

Illustration 35: The Heart Center

The Heart Center is the only Center of strong motivating action that has no direct access to the **Sacral Center**; it is a Center that is active in the **World of Biological Form** and functions primarily in your body chemistry and through your autonomic nervous system. Although it has access to conscious modulation, it operates automatically beneath any level of your awareness, functioning most of the time on the unconscious biological level. The Heart Center is also a Center of Ego and Will Power. It gives you Dominion (control) over yourself and your life. In addition, it controls the inflow of the resource Oxygen into your heart. When your Heart Center is colored in on your chart, your intention to take charge and to manage resources for yourself and others is strong; it may lead to promising more than you can deliver. When your Heart Center is uncolored you may experience challenges because you feel what other people want, and you may want to take care of their wants over your own. You may promise more than is right for you and defer your "will" to them, i.e., you may put their wants first and defer your sense of self-determination to them. In cases like this you may promise more than you can deliver and then may struggle internally because you feel internal pressure that may undermine your sense of your own value.

The double Trigram **Hexagram 51 (Thunder/Thunder) "The Arousing"** is in the astrological sign of **Aries (♈)**. Hexagram 51 is keynoted "Shock". When you deeply meditate on the flow of energy in the body map and its physiological components, you might surmise it sometimes is the shock that comes with the courage and strength that we mobilize when we face our fears. Sometimes we need to be "shocked" into action, otherwise, we may remain dissatisfied or unhappy in regard to our life situations.

Step 4: Understanding the Five Types

Noble Sciences teaches that there are five distinct "Types" or "Ways of Being" in humans that must be considered within the system. I validated these five Types as a statistically reliable measurement significant in describing how individuals function. My research and findings can be viewed at www.unifiedlifesciences.com.

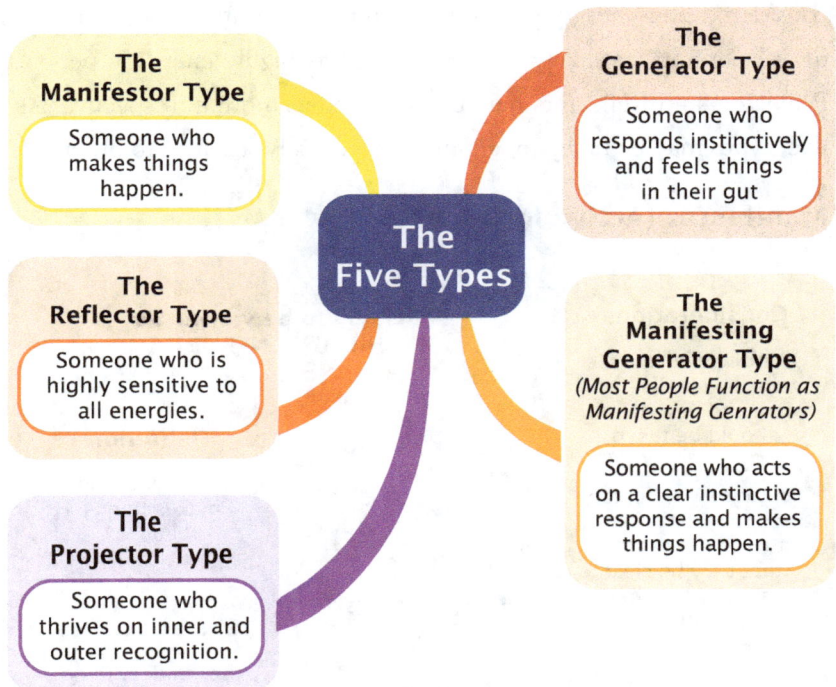

Illustration 36: The Five Types or Five Ways of Being
http://beyondhumandesign.com/other-tools/#the-five-types

The Five Types or Five Ways of Being
- The Manifestor (Active Initiator)
- The Generator (Instinctive Responder)
- The Manifesting Generator (Anticipatory Responding Initiator)
- The Projector (Self-Recognition of Inner and Outer Perceptions)
- The Reflector (Open Sensitivity)

Energy and Non-Energy Types

These five Types represent the way you face and meet the world. The Manifestor, the Generator, and the Manifesting Generator are considered energy Types while the Projector and the Reflector are considered non-energy types. Manifestors, Generators and Manifesting Generators are likely to be perceived as highly energetic people. Projectors and Reflectors may be perceived as more introspective. All five Types may be highly open and energetically sensitive to others.

Manifestor (Active Initiator)

The Manifestor (M) Type

Likely to be viewed as HIGHLY ENERGETIC

Considerations
- **Noble Sciences™** tells the Manifestor when and how action should be taken
- Because Manifestor likes to initiate things, it is very important that they be clear about what they are initiating.

Thoughts
- It is necessary for everyone to be able to initiate things for themselves as well as to allow receptivity when appropriate
- Manifestors should be certain that their actions are not premature or impulsive.

Definition
- Someone who makes things happen.

Characteristics
- They initiate things and help make them happen.
- They feel actively energized when they reach their goals.
- When they act on impulse or precipitously, they can cause turmoil.
- They tell other people about plans in order to prevent surprise and resistance from happening.

Illustration 37: Manifestor
http://beyondhumandesign.com/other-tools/#the-manifestor

A Manifestor gets things done because they're natural initiators. If you are a Manifestor, you can make things happen. You are active and feel enlivened when you reach your goals. You take action when you have clarity. However, when you act precipitously or impulsively, sometimes you cause turmoil. As a Manifestor, when you do not tell other people what you plan to do, they may be surprised and resist your wants and desires. Although only 8% of people are defined as Manifestors in their Mental, i.e., waking reality (solar) chart, you most likely function as a Manifestor in some way or another in your life. You are certainly capable of initiating some actions otherwise you wouldn't be able to function in society. Noble Sciences tells when and how you are most likely to take the best decisive and successful actions.

Generator (Instinctive Responder)

Likely to be viewed as ENERGETIC

Definition
- Someone who responds instinctively and feels things in their gut
- Someone who waits for inner clarity of response

Considerations
- Generators often find themselves very busy but they may just be spinning their wheels.
- Indecision may be the result of conflict between instinctive responses and what the mind interprets as a "should".
- **Noble Sciences™** teaches them to recognize the difference between instinctive responses and mental reactions.

The Generator (G) Type

Characteristics
- A Generator responds to things instinctively and then must use their human intelligence to assess what they feel and how they want to proceed.
- Asking a Generator questions allows them to tune into their gut level responses.
- If feelings are not clear, Generators must wait for clarity.
- Lack of clarity about feelings keeps a Generator doubtful about taking any path, leaving them in a 'stuck' state.
- Conflict between the mind and the gut may result in frustration.

Thoughts
- When Generators know clearly what they want instinctively and it agrees with their mental concepts, Generators may feel empowered.
- When in conflict or uncertain about what they feel, Generators may feel thwarted and confused.

Illustration 38: Generator
http://beyondhumandesign.com/other-tools/#the-generator

A Generator responds to things instinctively and then uses human intelligence to assess what they're feeling and the best way to proceed for gratification. As a Generator, if your feelings are unclear or you feel doubt, it is ideal for you to wait until you feel inner clarity. At times, indecision about what you feel may keep you, as a Generator, from taking any path. Furthermore, frustration may arise when inner clarity feels elusive. Often, you may find yourself busy, but you may be just spinning your wheels. Your indecision may be the result of conflict between your instinctive responses and what your mind interprets as a "should". Yet, when you know clearly what you instinctively want, and it agrees with your mental concepts, you are likely to feel empowered. When you are in conflict or uncertain about what you feel, you may imagine that you are thwarted and find yourself confused and frustrated.

Manifesting Generator
(Anticipatory Responding Initiator)

Considerations

Noble Sciences™ teaches them to recognize the difference between instinctive responses and mental reactions.

If upon envisioning their path, they still feel doubtful and confused or they still wonder about choices and responses, they must reconsider.

After envisioning a path, if it still feels congruent to their inner Self, they may proceed.

Responses, action, and effects must be congruent for them to feel at ease.

Likely to be viewed as HIGHLY ENERGETIC

Definition

Someone who acts on a clear instinctive response and makes things happen.

Characteristics

A Manifesting Generator uses a complex strategy in life and instinctively responds to things and initiates action to make things happen the way they want.

For a Manifesting Generator, instinctive responses must be clear.

Once clear, the Manifesting Generator must envision how those responses will play out in their world.

It is important for a Manifesting Generator to keep others informed so as to avoid miscommunication and resistance.

The Manifesting Generator (MG) Type

Thoughts

They always try to balance the Inner and Outer Self — the internal instinctive responses and the cognitive transformation or modification of those responses that shape their life.

They strive to become the most that they can be through the learning process. They are here to learn and to grow.

It's natural for them to take charge or to attempt to take charge of things.

Illustration 39: Manifesting Generator
http://beyondhumandesign.com/other-tools/#the-manifesting-generator

As a Manifesting Generator you use a complex strategy in life and apply both instinctive responses to things and active initiating to make things happen in the way you ideally want to have it happen in your life. First, instinctive responses must be clear, and then you must envision how your responses will play out in your world. If you consider the path you plan and it feels congruent to your inner Self, proceed. However, if upon envisioning your path, you feel doubtful and confused, you may wish to reconsider your path to maximize your probability of success.

As a Manifesting Generator you are here to learn and to grow. Part of the learning process is to become the most you can be. You always attempt to balance your Inner Self and Outer Self, your internal instinctive responses and cognitive transformation and you modify your responses to shape your life. Finding balance is part of the complex process of living. You are likely to be a Manifesting Generator in integrated functioning (the different composites in combined charts); at those times, you know and feel at ease within your Self, meaning when looking at the charts separately, you may not always appear to function in this manner. It's natural for you to take charge or attempt to take charge of things. You personify empowerment, i.e., human potential being lived fully. At least 95% of people in functional life become Manifesting Generators and rely upon this energy flow in how they make decisions to live an aligned life.

Projector
(Self-Recognition of Inner and Outer Perceptions)

The Projector (P) Type

Likely to be viewed as INTROSPECTIVE

Considerations

- Projectors learn through their contact with different energy types and distinct energy configurations in people and how those energies affect them.

- When they feel validated by another person, they gain energy or feel momentum toward continuing their process rather than denying it.

- Communication is key for Projectors.

Definition

Someone who thrives on inner and outer recognition.

Characteristics

- A Projector feels alive when they connect to people or to their inner Self.

- A projector reveals their qualities and strengths.

- Because a Projector is not designed to take action, they might not be aware of their gut feelings.

- The projector benefits when another person acknowledges what is going on beneath the surface.

- Projectors may be highly sensitive to acceptance or rejection.

Thoughts

- **Noble Sciences™** teaches Projector to move beyond the limitations of their own Waking Self and its personality. They become more fully able to move in the universe with non-attachment and freedom. This transformation into limitless possibilities is everyone's potential and their birthright.

- The Projector often needs to turn energy inward to gain clarity.

Illustration 40: Projector
http://beyondhumandesign.com/other-tools/#the-projector

As a Projector you feel an inner stirring of energy that is ready to connect with some aspect of your being when you are properly recognized and acknowledged. As a Projector you need to turn energy inward to gain clarity. The Projector parts of you benefit when another person, or situation, invites or acknowledges what's going on beneath the surface of your conscious awareness. You are likely to feel validated when another person recognizes your sensibilities and abilities and you gain energy or feel momentum toward continuing this process rather than denying it at those times. However, as a Projector, you function optimally when you are "tuned in" to what you perceive and experience and recognize your own Self. It is this kind of self-awareness and self-knowledge that validates you and builds self-esteem and inner resilience.

Reflector (Open Sensitivity)

Highly RECEPTIVE to ENERGY and INTROSPECTIVE

The Reflector (R) Type

Considerations

- **Noble Sciences™** aides them in getting in touch with their open nature so they can feel empowered thus they avoid feeling overwhelmed, exhausted and confused.

- The strength for a Reflector is in their open awareness because they relfect their true nature rather than trying to live as one of the other Types.

- Keeping options open helps the Reflector

Definition

Someone who is highly sensitive to all energies.

Characteristics

- Reflectors receive information from cosmic forces and from other beings openly with few predetermined filters.

- Very few individuals remain the Reflector Type in their integrated functioning.

- Reflectors are totally vulnerable in the sense that every one of their Centers is open.

- Reflectors are sensitive to people, places, and things. They experience and reflect their sensitivities.

- Being open and vulnerable to outside conditioning can be difficult for Reflectors.

Thoughts

- During sleep open receptivity integrates in the Self through different configuration so neural pathways. Each individual expresses a unique view or perspective on their own way of being.

- Changing conditions open new possibilities for the Reflector Type.

- During sleep most individuals become Reflectors. During this phase of consciousness they move into the cosmic soup in which all humanity merges in consciousness as one being.

Illustration 41: Reflector
http://beyondhumandesign.com/other-tools/#the-reflector

As a Reflector you receive information from cosmic forces and from other beings openly with few predetermined filters. You are highly sensitive to all that is around you on all levels. When in a positive state, you experience "being in the flow." When in a negative state, the Reflector part of you may feel overwhelmed by information and thus, powerless. During sleep you are likely to function as a Reflector. During this phase of your consciousness you move into the "cosmic soup" in which all humanity merges their consciousness as one being in unity. You are unlikely to remain a Reflector Type in your integrated functioning charts, as your composites may reveal that you express yourself through other Ways of Being at different times, depending on planetary movements..

By understanding your own multi-dimensional charts and which centers are switching on and off, as well as noting the charts and configurations of the people you're relating to, insight can be gained into how your patterns impact your own consciousness as well as how the patterns of others impacts your emotional reactions. Charts never remain constant in their defined Type over any six-month period. They rather reflect the breathing living organism of the cosmos of which you are a perfect reflection within yourself. Ultimately, Noble Sciences helps realign you with your true self and how you naturally make decisions in order to navigate the world from within the flow. By knowing one, you know all. By knowing all, you know one. Know yourself. You are the universe.

An Example Chart - Steve Jobs

The power of Noble Sciences work can best be experienced by having a reading by me, and by learning to read your own chart. To give you an example of a chart and what is involved in reading one, take a look at the Steve Jobs chart and the transcript of the short reading I did on his chart. The reading follows this overview table; all charts follow the reading.

Steve Jobs
February 24, 1955 @ 19:15 PST (08h GMT), San Francisco, USCA

How you Function – Total Integrated Manifesting Life:
Manifesting Generator (Emotional)
Always envision how you protect things playing out prior to taking action or making commitments.

Illustration 42: Steve Jobs' Birth Data

Chart	(Prenatal/Natal) Activation/Recognition Field Approach to Life	
	Type	Description
Composite	Generator (Emotional)	Wait for clarity in your internal responses.
Mental	Generator (Emotional)	Pay attention to thoughts about your responses
Spiritual	Generator (Emotional)	Wait until you have a deep inner knowing of your responses.
Emotional	Projector	Notice if your feelings serve your goals and enhance self
Physical	Projector	Recognition of physical limitations Builds strength.

Illustration 43: Steve Jobs' Prenatal Overview

Chart	(Prenatal/PostNatal) Activation/Recognition Field Approach to Life	
	Type	Description
Composite	Manifesting Generator (Emotional)	Right action moves forward with clear actions that mobilizes support internally and externally.
Mental	Manifesting Generator	Acting from a place of inner clarity requires awareness of the needs of others as well as of yourself.
Spiritual	Generator	Continued certainty about your decisions despite slight nuances of adjustment to them, honor your inner process and support it.
Emotional	Projector (Emotional)	New options for re-framing perceptions opens up a new range of feelings.
Physical	Projector (Emotional)	Influences from others on health require alertness to your body chemistry that involves self-discipline.

Illustration 44: Steve Jobs' Postnatal Overview

Worlds:

Composite (Prenatal/Natal) Recognition Field: Approach to Life.
Mental/Interactive Reality, World of Earth
Spiritual/Archetypal Sleep & Dream World
Emotional/Inspirational Angelic World
Physical/Hormonal Biological World
Total Integrated – Manifesting Life
Composite (Natal/Post) Activated Integrated – Response to Life

Type Definitions:

Manifesting Generator – Anticipatory Responding Initiator
Generator – Instinctive Responder
Manifestor – Active Initiator
Projector – Self-Recognition of Inner and Outer Perceptions

Illustration 45: Steve Jobs' Key Worlds

Steve Jobs was born on February 24, 1955, at 19:15, Pacific Standard Time. In looking at his chart, from the multidimensional Noble Sciences perspective, you'll notice that there's a significant change between the different layers of personality, and functioning consciousness, revealing Steve as an Emotional Generator, with the 41st Gate activated by the pre-natal Moon and the natal Chiron. In addition, the 61st Gate activated both Chiron and Venus, so there was a strong drive for Steve to understand the inner mysteries of life. In addition, he had a visionary perspective that came from a deep place within him self, which included a deep understanding of the needs of the collective in terms of innovation. The chart shows a person who knew the value that creative innovation and its resulting awareness could bring to the planet.

His visionary nature was very focused and very deep, coming from a core instinct that tapped into the collective. This visionary instinct is confirmed in the spiritual chart with the 50th Gate, activated by Neptune in all the charts. A week after birth there was a connection between Gate 50 and Gate 27, Gates that connect the Sacral Center to the Splenic Center. This activation narrowed Steve's focus connecting his value of himself as a spiritual being who cared deeply from his heart, with the need to align energies from a higher perspective and bring the minutest detail through in the way he was able to realize the purpose of his Self.

On a deep level of core being, Steve only felt truly fulfilled and gratified when he was able to create and manifest his visions on the physical level. This is illustrated in Steve's composite chart, revealing that he was an extremely strong Manifesting Generator, with every center colored in. In fact, it's clear from his composite chart that not only was Steve's total integrated being a very strong Manifesting Generator, but he had strong visionary goals and really valued how much he cared about what he created and shared with the world.

On the inner plane, Steve was extremely open, very sensitive, and very vulnerable to the energies around him. He picked up on the collective energies that were around and deeply absorbed them, especially on a physical level. Moreover, he was also greatly affected by anything that interfered with his need to be caring on a collective level. He had strong emotional reactions when this occurred, and others then took in this emotional energy, and reacted to it in their own way. He was a sensitive person who was finding his own role in the world. He used his creativity to contribute his understanding of human needs with a sense that he had higher vision and purpose. He wanted to empower people through aligning structure and function of objects that would make life easier. The 12th Gate was active in the postnatal solar chart where there was the tribal Channel 40/37, connecting to his solar plexus in the postnatal solar minute chart, i.e., the emotional chart. There are a lot of complex connections that turn on and off in a very dynamic way in this chart.

The 12th Gate connects to the 22nd Gate connecting the pathway to the throat. When you trace the patterns in the chart, you'll see that Steve's goals connected with his values and empowered the role that he played in life, as well as how his vision was verbalized or expressed in the way that touched the broader society as a whole. When his energy flow was interfered with, there was an emotional reaction that then set off a domino effect on the people around him.

In Steve's chart several planetary positions trigger hexagram placements in the body graphs that point to his being a visionary pioneer. Steve's Chiron is two degrees, nineteen minutes of Aquarius (2°♒19') in his fifth astrological house (the house natal to Leo), placing Chiron by sign and degree in Gate 41, Hexagram 41 in the Body Graph. It activates the root center. This focuses creative energy for utilizing inner and outer resources in service of stability and efficiency. In Aquarius, Gate 13 is also activated by Mercury, the messenger. Gate 13 defines fellowship and points to a capacity to activate a group of colleagues toward a common and higher goal. His Mercury is in his fifth astrological house natally, indicating and strengthening the creative expression of the 13th Hexagram.

Steve Jobs came into the world with a strong collective purpose. His chart shows that his openness and connection with the Ajna Center (third eye or intuition) and his throat center made him a channeling vehicle through the Channel 17/62, creating his ability to focus on detail and use his unique intelligence. Coupled with the 43rd Gate and the 56th Gate, his storyteller nature came through, expressing itself in a very deep and very powerful way. He knew what needed to happen and how to translate it for the collective. He brought conscious awareness that has changed the lives of many people in significant ways by making technology intuitive and user friendly.

The Hexagram activations in Steve's chart tell a very significant story in terms of their planetary and Gate activations. Gate 58, in the astrological sign Capricorn, located at the root center, points toward the splenic center. Gate 52, in the astrological sign of Cancer, is activated by the north node in Steve's prenatal Solar Chart and in both his Prenatal and Postnatal Spiritual Charts. These placements give a strong oppositional energy that activated Steve's finding balance so he knew what was energetically aligned and conscious.

In his postnatal Spiritual chart the 52nd Gate is activated by his Moon, giving him a strong emotional confirmation when aligned energies are "right" in terms of his consciousness. Gate 52 is in the root center pointing toward the sacral center and defining a natal Channel to the sacral center through Gate 9, in the astrological sign of Sagittarius. Gate 9, a Gate of focused energy that connects with the inner stillness of the 52nd Gate, is activated by his prenatal solar Sun. Gate 10, activated astrologically by the north node in Capricorn, in his self center, and Gate 15, activated by the south node astrologically in Cancer, also in the Self Center, impel him to take on a role of great responsibility for bringing consciousness to his life activities and purpose. Because Gate 15 is a bridge Gate to all four worlds, his sense of what needed to happen to bring his values (his Channel 50/27 Neptune and Mars in his postnatal Spiritual chart) to manifestation defined his long term goals (Channel 34/20/10 and Channel 54/32) in his composite chart.

Noble Sciences charts demonstrates the importance of a multidimensional perspective in Steve Jobs charts. In a basic Human Design chart much of the nuance and process of how he functioned would be lost and his great contribution and his potential to manifest in the world would have been overlooked. In his Human Design chart, he would have been classified as a split definition generator. In Noble Sciences charts, which go beyond Human Design, Mr. Jobs is seen as a powerful Manifesting Generator who lived with a deep commitment to spiritual wisdom and who needed time to process his ideas and set goals that would honor his values and sense of aligned energy that could bridge dimensions and manifest in a loving and integrated way.

78 ♦ COSMIC SECRETS

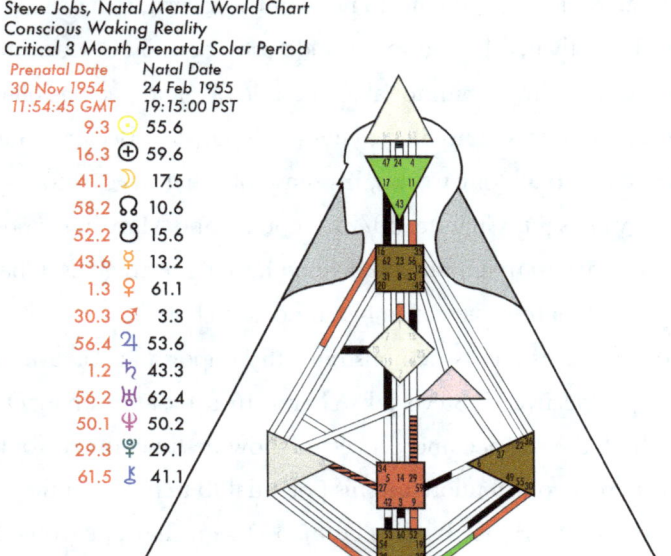

Steve Jobs, Natal Mental World Chart
Conscious Waking Reality
Critical 3 Month Prenatal Solar Period

Prenatal Date	Natal Date
30 Nov 1954	24 Feb 1955
11:54:45 GMT	19:15:00 PST

9.3	☉	55.6
16.3	⊕	59.6
41.1	☽	17.5
58.2	☊	10.6
52.2	☋	15.6
43.6	☿	13.2
1.3	♀	61.1
30.3	♂	3.3
56.4	♃	53.6
1.2	♄	43.3
56.2	♅	62.4
50.1	♆	50.2
29.3	♇	29.1
61.5	⚷	41.1

55-59/9-16: The Changing Pattern of Spirit

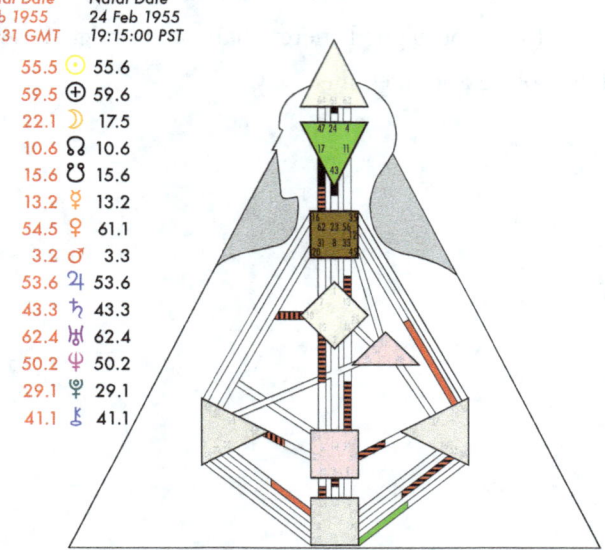

Steve Jobs' Prenatal Emotional World Chart
Emotional Underlay
Critical 2 Day Prenatal Solar Minute Period

Prenatal Date	Natal Date
23 Feb 1955	24 Feb 1955
16:16:31 GMT	19:15:00 PST

55.5	☉	55.6
59.5	⊕	59.6
22.1	☽	17.5
10.6	☊	10.6
15.6	☋	15.6
13.2	☿	13.2
54.5	♀	61.1
3.2	♂	3.3
53.6	♃	53.6
43.3	♄	43.3
62.4	♅	62.4
50.2	♆	50.2
29.1	♇	29.1
41.1	⚷	41.1

55-59/55-59: The Changing Pattern of Spirit

Illustration 46: Steve Jobs' Prenatal Solar Charts

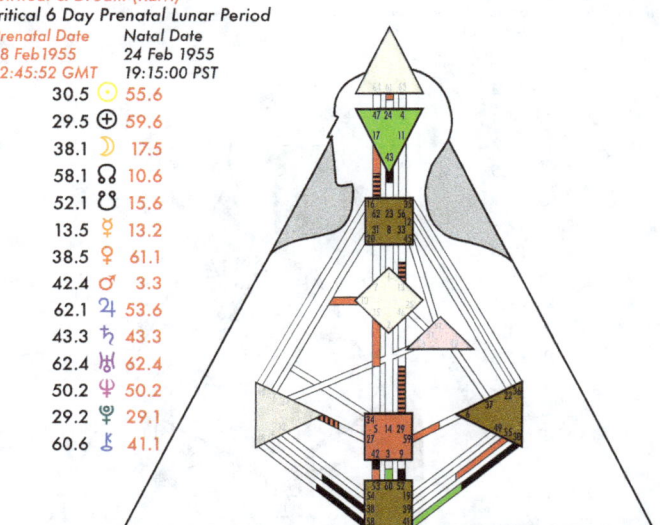

17-18/38-39: The Changing Pattern of Upheaval (Turmoil)

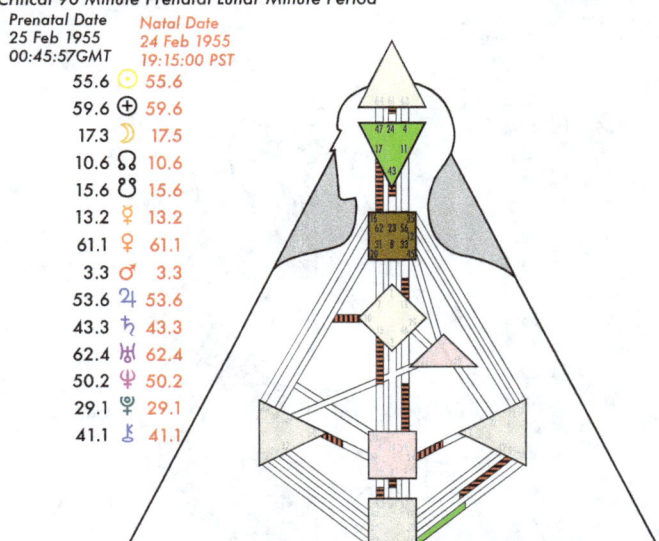

17-18/17-18: The Changing Pattern of Upheaval (Turmoil)

Illustration 47: Steve Jobs' Prenatal Lunar Charts

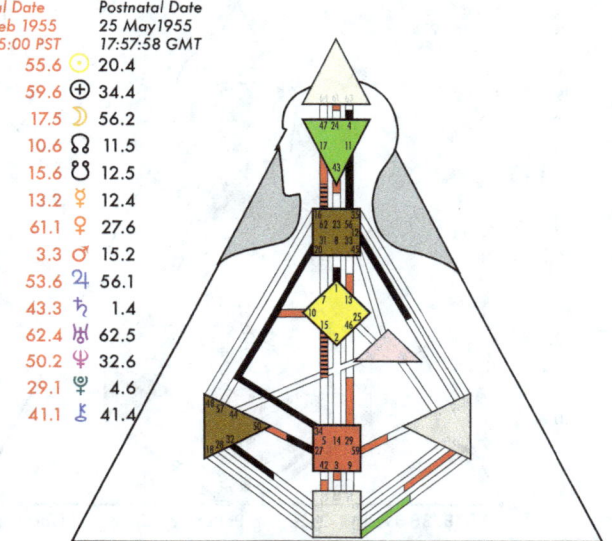

20-34/55-59: The Foundation Pattern of the Sleeping Phoenix-2

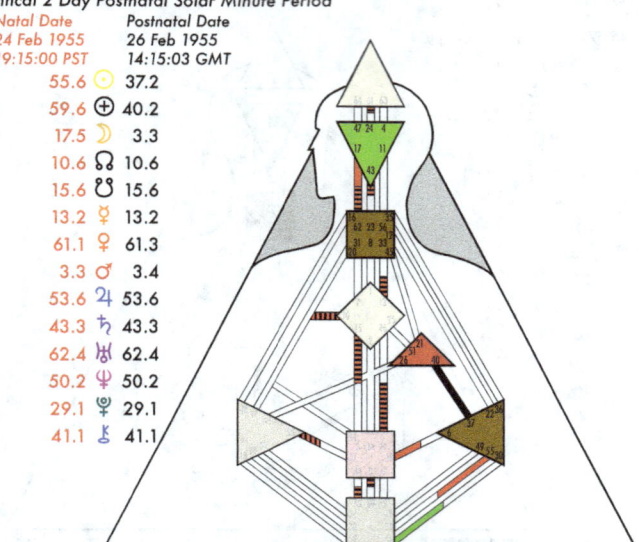

37-40/55-59: The Foundation Pattern of Planning

Illustration 48: Steve Jobs' Postnatal Solar Charts

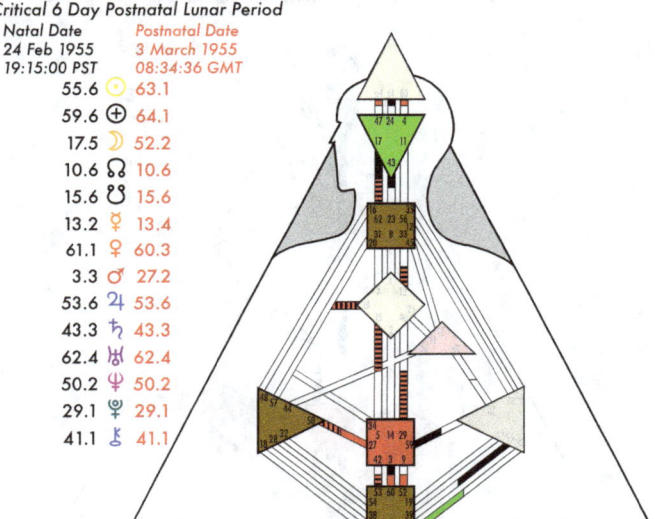

52-58/17-18: The Foundation Pattern of Service-2

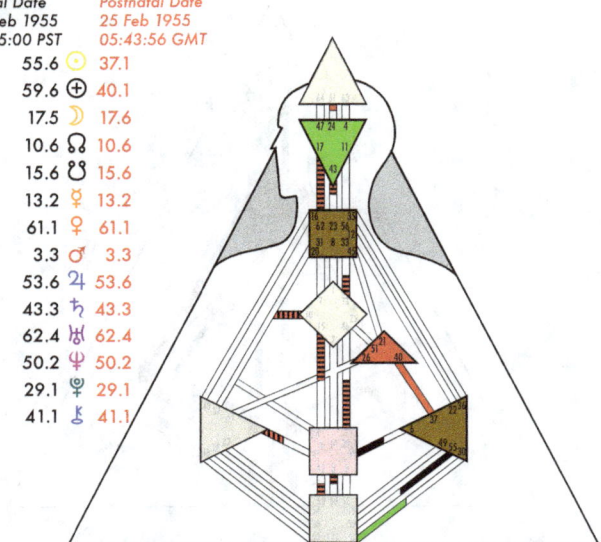

17-18/17-18: The Changing Pattern of Upheaval (Turmoil)

Illustration 49: Steve Jobs' Postnatal Lunar Charts

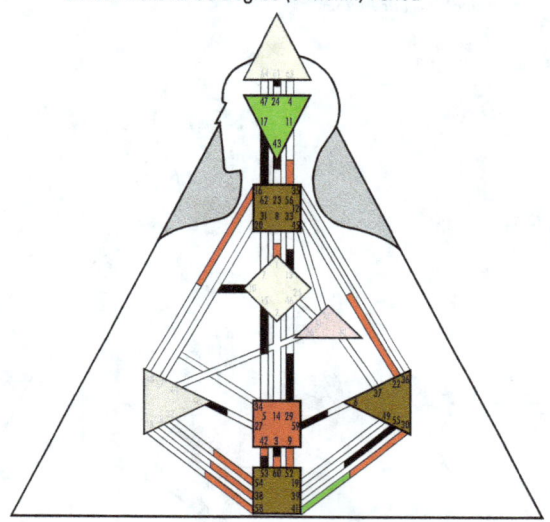

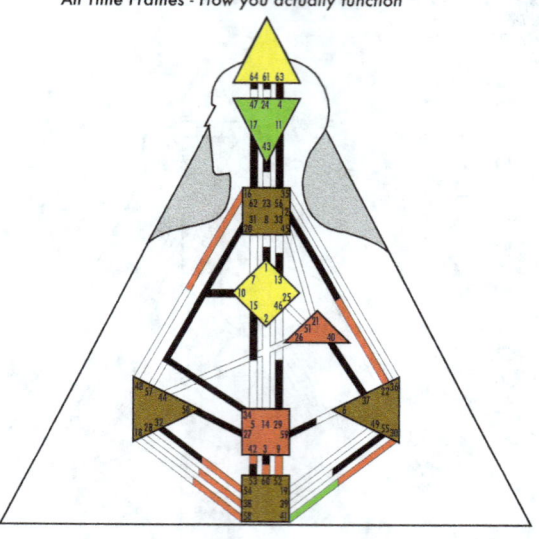

Illustration 50: Steve Jobs' Composite Charts

UNLOCKING NOBLE SCIENCES SACRED SYNTHESIS ♦ 83

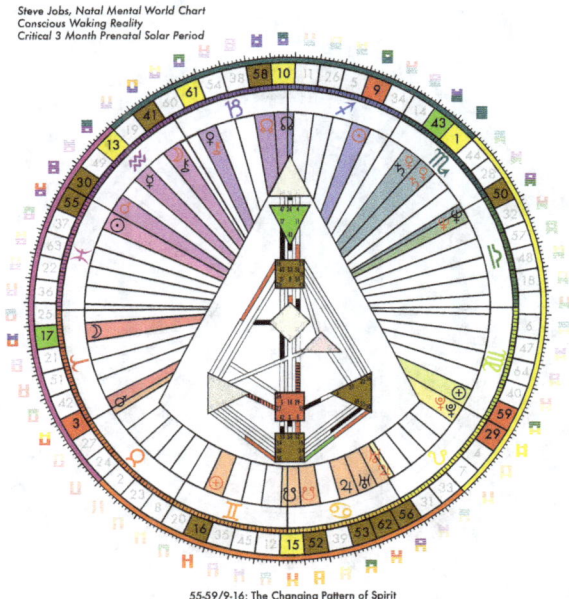

Illustration 51: Steve Jobs' Mandala Prenatal Solar

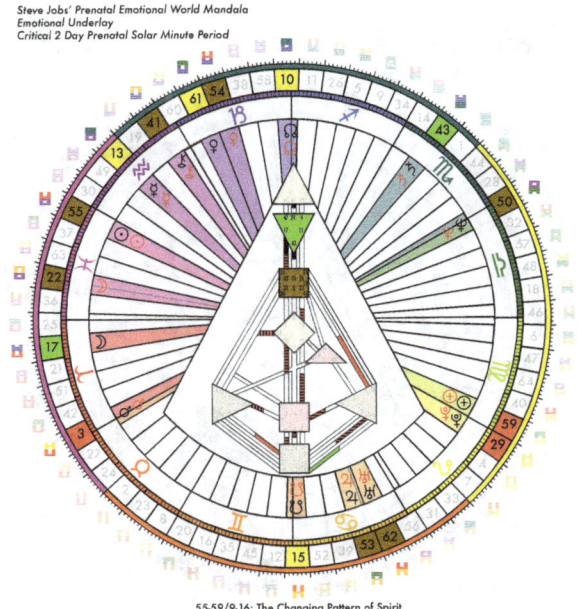

Illustration 52: Steve Jobs' Mandala Prenatal Solar Minute

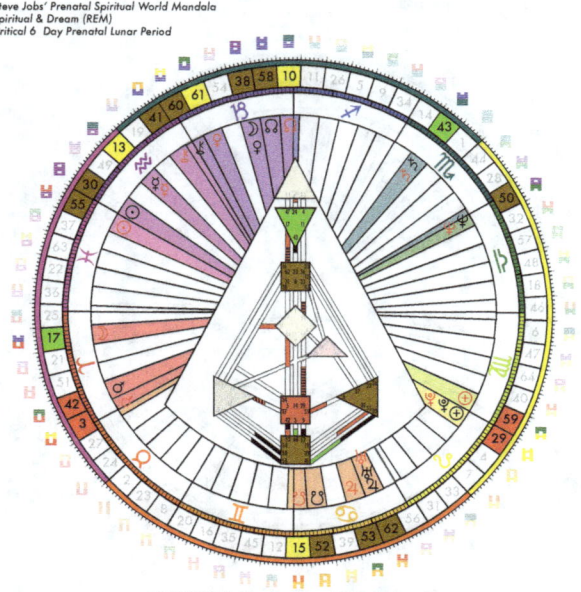

Illustration 53: Steve Jobs' Mandala Prenatal Lunar

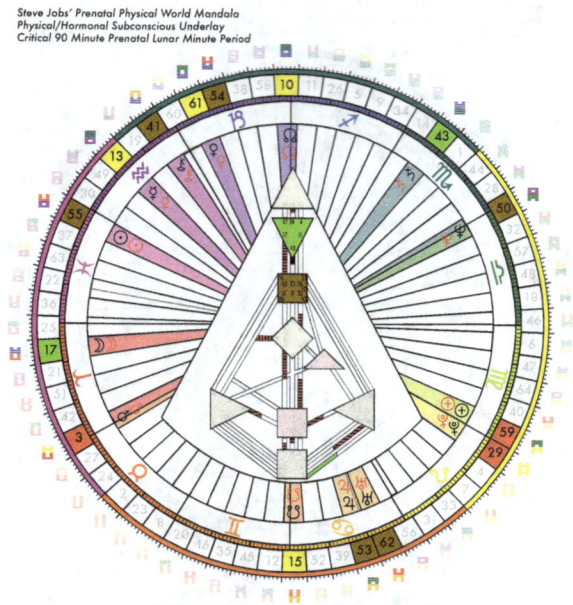

Illustration 54: Steve Jobs' Mandala Prenatal Lunar Minute

UNLOCKING NOBLE SCIENCES SACRED SYNTHESIS ♦ 85

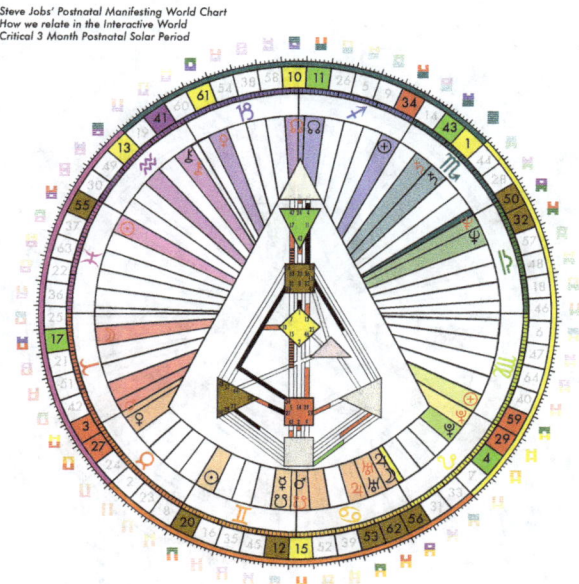

20-34/55-59: The Foundation Pattern of the Sleeping Phoenix-2

Illustration 55: Steve Jobs' Mandala Postnatal Solar

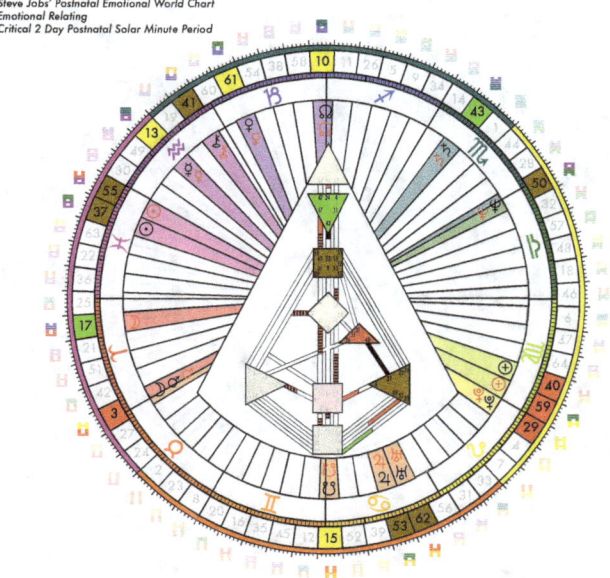

37-40/55-59: The Foundation Pattern of Planning

Illustration 56: Steve Jobs' Mandala Postnatal Solar Minute

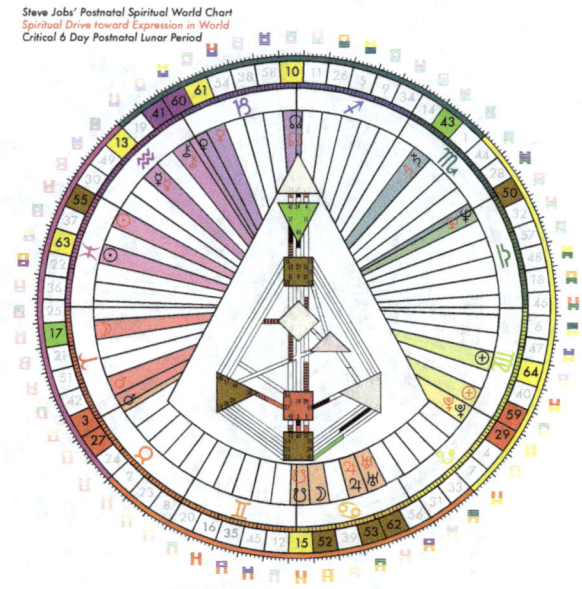

Illustration 57: Steve Jobs' Mandala Postnatal Lunar

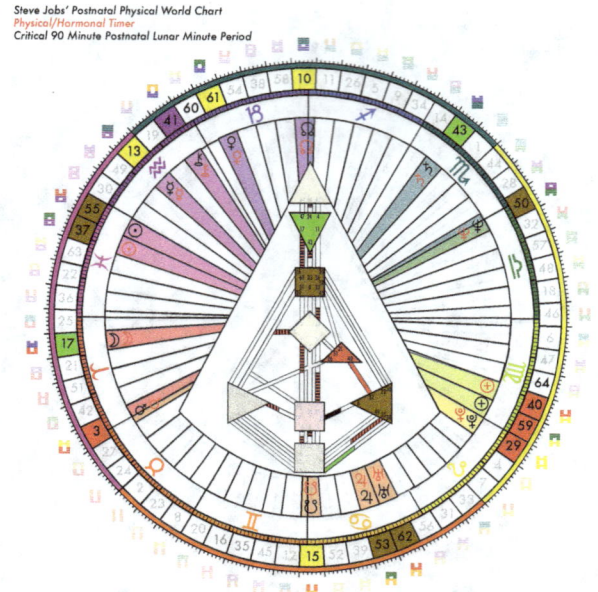

Illustration 58: Steve Jobs' Mandala Postnatal Lunar Minute

Enhancing Practices

Noble Sciences offers practical actions or enhancing practices that assist you in aligning with the flow of cosmic energy. Below is an example of one suggested practice that was particularly useful as we entered a period that called for inner reflection.

Mind Maps

Because the information in Noble Sciences synthesizes many technically complex wisdom traditions and adds deepening layers of complexity and multidimensionality, I often organize the information presented into Mind Maps. The Charts of the Five Types and the Illustration of Journaling are all Mind Maps. A Mind Map is a graphic representation of concepts, words, and other information arranged in a circle around a central key word or idea. The use of Mind Maps dates back to the 3rd century. They help people make sense of otherwise complex written material and allow them to process information through graphics. They reduce cognitive load or information overload. Thus, mind maps improve and enhance recall and learning. In recent years, with the emphasis in education on whole brain learning and accelerated information processing, Mind Maps are increasingly and more effectively being used.

Journaling: An Integrative Tool that Enhances your Self

Illustration 59: Journaling Mind Map
http://beyondhumandesign.com/?s=journaling

Key to Colors and Symbols

Sign Name and Glyph		Planet Name and Glyph		Center Color/Shape Center Name	Trigram	Keynote
Aries	♈	Mars	♂	Crown (Head) ◇	☰	Thunder
Taurus	♉	Venus	♀	Ajna ▽	☷	Earth
Gemini	♊	Mercury	☿	Throat ▢	☶	Mountain
Cancer	♋	Moon	☽	Self ◇	☴	Wind
Leo	♌	Sun	☉	Heart ◁	☳	Heaven
Virgo	♍	Mercury	☿	Sacral ▢	☱	Lake
Libra	♎	Venus	♀	Splenic △	☵	Fire
Scorpio	♏	Pluto	♇	Solar Plexus ▽		
Sagittarius	♐	Jupiter	♃	Root ▢		
Capricorn	♑	Saturn	♄	**Body Graph Colors**		
Aquarius	♒	Uranus	♅	Mental/Interactive Reality Earth World (yellow)		
Pisces	♓	Neptune	♆	Spiritual/Archetypal Sleep/Dream World (purple)		
		Chiron	⚷	Emotional Mystical Inspiration/Angelic World (orange)		
		Earth	⊕	Physical/Hormonal Circadian Biological World (red)		
		North Node	☊	Active Gate (black)		
		South Node	☋	Receptive Gate (white)		
				Chiron Activated Gate (green)		

Types: Manifestor – Generator – Manifesting Generator – Projector – Reflector

Illustration 60: Color Key

90 ♦ COSMIC SECRETS

♈ Aries	♉ Taurus	♊ Gemini
♋ Cancer	♌ Leo	♍ Virgo
♎ Libra	♏ Scorpio	♐ Sagittarius
♑ Capricorn	♒ Aquarius	♓ Pisces

☉ Sun	☽ Moon	☿ Mercury
♀ Venus	♂ Mars	♃ Jupiter
♄ Saturn	♅ Uranus	♆ Neptune
♇ Pluto	☊ North Node	☋ South Node
⚷ Chiron	⚴ Pallas	⚵ Juno
⚳ Ceres	⚶ Vesta	AS Ascendant
MC Midheaven	⊗ Part of Fortune	⊕ Earth

☌ Conjunctions	☍ Oppositions	△ Trines
□ Squares	⚺ Semi-Sextiles	♃ Sextiles
⚻ Inconjuncts	∠ Semi-Squares	⚼ Sesquiquadrates
Q Quintiles	✶ Septiles	⋈ Noviles
℺ Quindeciles	// Parallels	⋕ Contra-Parallels

Illustration 60: Astrology Legend

UNLOCKING NOBLE SCIENCES SACRED SYNTHESIS ♦ 91

Planet	Time to Orbit Sun	Natal Return Cycle	Time in a Gate	Time in a Line
☉ Sun	365.25 days about 1° per day	Yearly About 1°/day	5.7 days	.95 days
⊕ Earth	365.25 days about 1° per day	Yearly About 1°/day	5.7 Days	.95 Days
☽ Moon	27.3 days about 12°/day	Lunar return is monthly	.46 days	.08 day
☿ Mercury	87.96 days (always within 28° of the ☉ up to 2°30'/day	About 3 months	1.4 days	.23 day
♀ Venus	224.68 days 1°15'/day	About 7.5 months	3.5 days	.59 day
♂ Mars	686.95 days (1.89 years) 0°40'/day	About 23 months	10.78 days	1.8 days
♃ Jupiter	11.87 years about 30°/year	11.87 years	67.69 days 2.25 months	11.3 days
♄ Saturn	29.46 years about 1°/year	29.46 years	168 days 5.6 months .46 year	28.03 days .08 year
♅ Uranus	84.05 years about 4°/year	84.05 years	479.34 days 15.98 months 1.31 years	79.92 days .22 year
♆ Neptune	164.81 years about 1° to 2°/year	164.81 years	940 days 31.33 months 2.58 years	156.77 days .43 year
♇ Pluto	248.54 years about 1°/year Elliptical orbit not equal in all signs	248.54 years	1417.45 days 47.25 Months 3.88 years	235.6 days .65 year
⚷ Chiron	50.7 years	50.7 years	289.15 days 9.64 months .79 days	48.19 days .13 years
☊ North Node	18.6 years about 20°/year retrograde	About 18.6 years	106.08 days 3.54 months .29 years	17.68 days .05 year

(Mean planetary orbits from Astronomy: From the Earth to the Universe. 3rd Edition. Jay Pasachoff. Sunders, NY. 1987.)

Illustration 61: Table of Planetary Movement

0° Zodiac Position of Astrological Sign	Zodiac Symbol	Planetary Ruler by Astrological Sign	Hexagram & Line Number at 0° of Sign	Hexagram Glyph
Aries	♈	♂	25.2	25
Taurus	♉	♀	3.4	3
Gemini	♊	☿	8.6	8
Cancer	♋	☽	15.2	15
Leo	♌	☉	56.4	56
Virgo	♍	☿	29.6	29
Libra	♎	♀	46.2	46
Scorpio	♏	♇	50.4	50
Sagittarius	♐	♃	14.6	14
Capricorn	♑	♄	10.2	10
Aquarius	♒	♅	60.4	60
Pisces	♓	♆	30.6	30

Illustration 62: Hexagram Correspondences
to Zero Degree Points of Each Zodiac Sign

Legend: Astrological Sign is Color Coded by its Sign at 0° point;
Zodiac Symbol is Color Coded;
Planetary Ruler of Sign is Color Coded by Astrological Sign;
Hexagram Gate and Line is Color Coded by Color of Center in Body Graph;
Hexagram is Color Coded by Astrological Sign of
Lower Trigram of Self Center

Postscript

In my professional experience, individuals interested in cosmic knowledge also value self-knowledge. If you are reading this book, you are deeply committed to self-growth and to aligning your life to spiritual principles that serve your deepest self.

Many available tools can empower you in making conscious choices that will aid you in reaching your life goals. However, the first step in growing consciousness always requires honest self-discovery and self-knowledge. Powerful questions evoke powerful answers. Moreover, powerful questions activate deep changes that bring your awareness toward fulfilling actions.

Consider the following questions:
- What do you want to have happen now in your life?
- What do you love about your life?
- What challenges are you facing?
- What do you want to change in your life?
- What 3 things do you want to do in the next 30 days?
- What dream drives you?

Research shows that limiting beliefs can keep you from realizing your goals and dreams. Everyone needs supportive nurturing and loving guidance to intentionally change their limiting beliefs with minimum stress. You can receive this support and soar to your greatest heights. From my work since 1974 with individuals and groups in private clinical practice, I can assure you that without delving into the contents or past events of your life, your dreams are within reach. What happened in your life in the past can inform you, but it need not lock you into old patterns. When we experience negative events, we code them non-verbally and metaphorically, and we get stuck in our feelings. Now we have powerful tools that can by-pass painful memories and allow you to remain free from the blocks created by painful experiences so you can reach success and live the life of your dreams. Through rewriting the story that took place, you'll be released from the powerful hold they have on you.

If one of your goals is to assure that you are on your right path, working with Noble Sciences can serve you well. The information coded into Noble Sciences Charts provides a roadmap of how you function in the world and reveals a way to create a strategy for the best timing and planning in order to manifest the things you want to achieve in your life.

Noble Sciences' life-changing coaching tools synthesize ancient wisdom traditions and rigorous social scientific research, and when recognized and used together, they activate knowledge from sacred disciplines that open and access information deep within you so your daily life and relationships can express your true nature.

The depth of my extensive professional knowledge in conjunction with the Noble Sciences system, utilizes the wisdom of the cosmos that reveals truths about you at the core of your being. Begin this exciting journey of self-discovery now.

- Make authentic conscious choices.
- Find your truelife-purpose.
- Find your soul mate.
- Improve your relationships.
- Experience abundance in all areas of your life.

Like most of us, you probably want to understand your place in the Universe, and although you've heard that you are not separate from it, your life may need some "tweaking" for you to fully experience, balance, and feel the law of attraction in operation.

Today, millions of people are returning to the wisdom of the ages to understand themselves and the world around them. Science validates the inner world, continually giving us updated knowledge to determine when intuition and interpretation may be off course.

Awareness Always Precedes Choice

- Learn how to adjust your life course.
- Make conscious choices.
- Take effective intentional action.
- Achieve your goals.
- Manifest the life you truly want.

Noble Sciences Sacred Synthesis blends the personal and the universal, the scientific and the mystical, the ancient and the modern. I've created tools that teach you how to consciously align with cosmic influences and I've used them over the years to help countless clients find the path that most feeds their spirit. Rather than feeling you're at the whim of your subconscious mind or life circumstances, you'll learn to understand how cosmic energy helps you to consciously exercise your self-will in order to live life to its fullest extent.

Understand and Transcend

Astrology is a helpful tool in this growth process because the moment of birth marks an important cosmic event in individual development. Important indicators of your personality are determined in a window of time that occurs three months before the day upon which you're born, to three months after that date. Because I read what is coded into the chart maps I use, I'm able to recognize patterns of response in your behavior that express timing and decision-making strategies, which are opportunities for growth. These strategies are key in discovering your life purpose and living a fulfilled life.

Your development is unique. A cookie-cutter approach that labels or puts you into a box does not work. Even though some basic types of energy characterize people and their responses, it is important to recognize that you are never locked into any specific response. Your beliefs affect your choices. Because you have awareness, your choices, when made consciously, can guide you toward positive action.

> **Noble Sciences Combines:**
>
> *Kabbala – Tree of Life*
> *Hindu Chakra System*
> *The Tao – I-Ching*
> *Astrology*
> *Human Design*
> *Developmental Psychology*

Noble Sciences Tools synthesize complex disciplines: the Tree of Life, Chakras, Astrology, the I-Ching, Human Design, and Human Development. The blueprint shown in the chart map of an individual is useful, but blueprints only guide the construction of a building. The builder's choices and ability to make adjustments in real-time bring the blueprint to life. You need not learn how to read the blueprint. I read the blueprint for you and tell you what you, the builder, need to know in order to make your life work better.

Many people ask, "Do cosmic influences predetermine my life?" The answer is that cosmic influences impel you; they do not compel you. For example, cosmic influences may be the reason you enjoy playing piano, but they do not force you to do so. The information used and interpreted in the Noble Sciences charts gives you the awareness and information you need to make informed choices about how you want to proceed with the rest of your life. You always have the ability to move beyond personal and cosmic limitations. With love and nurturing, limitations fall away so possibilities can become realities.

Navigating through life with Noble Sciences is a bit like using a weather report. You might experience comfortable weather much of the time, but without a weather report, you won't be prepared for the storms and extreme changes in temperature that will occur. Noble Sciences tools are like a weather report that tells you what cosmic forces impact you. Thus, you can choose when to stay in your comfort zone and when to stretch your self to reach your goals. Noble Sciences tools give you the knowledge to avoid discomfort and disappointments by anticipating what is up ahead for you.

As a Noble Sciences coach, I provide you with an "energy forecast" instead of a weather forecast. The tools in this discipline enable me to be an energy guide who facilitates your ability to recognize the context and climate of your life on a personal and cosmic level. You'll see what unfolds around you and gain greater ability to change your responses for the better, to know when to "grab your umbrella," and when to leave it behind.

Weather is a personal thing. Some people prefer 80 degrees and sunshine, others enjoy 65 degrees and a cool breeze. Just like the weather, each person has requirements, needs, and desires, and is affected by cosmic energy based on their own unique internal architecture. Learning about the depth of your unique complexity (consciousness) allows you to conclude for yourself what you want and how an "energy climate" is likely to impact you.

Tools

Noble Sciences provides you with the depth and guidance of a professional who has the vast knowledge and experience of esoteric wisdom that allows you to take your life to a whole new level. You'll learn crystal clear ways of finding the love of your life or starting a successful business and feel empowered to not only handle the stress of your day, but to evolve into the person you want to be. Through Noble Sciences, you will live the life you want to live because the choices you'll make will be based on divine wisdom and made consciously, bringing your deepest desires to life.

Noble Sciences Tools

- Treat you compassionately as a unique individual.
- Facilitate awareness of intentional choices making you a conscious decision maker and creator in your life.
- Teach you skills that expand your perspectives.
- Enhance your insights into how you function.
- Coach you with specific ways to approach all areas and aspects of your life.
- Help you set realistic, measureable, and achievable goals.
- Illustrate how to attune and navigate to the "energy climate" of the day.

Access valuable information through Noble Sciences Sacred Synthesis including

- Action points published daily on www.noblesciences.com and tweeted from https://twitter.com/ByndHumanDesign.
- Engaging software showing real-time images of energy in motion so you see planetary energy shifts as they happen
- Usable multi-sensory tools based on key whole-brain research
- Special updates on my research, coaching, and latest activities to keep you informed and facilitate your continual evolutionary process

I do the work. You reap the benefits.

This e-book took you through each of the key concepts of Noble Sciences. From background information, to the science behind the scenes, you learned about its ancient roots and its scientific underpinnings.

If you are interested in the most technical aspects of my research, you can find a wealth of information at Noble Sciences and the research I did at [Unified Life Sciences](). Unified Life Sciences was founded in 1999 to research Human Design as a complete system. Unified Life Sciences grew into Noble Sciences and dedicates itself to simplifying information for its audience based on clinical, coaching, and social scientific analysis. You only need the desire to learn about yourself to benefit from what Noble Sciences has to offer; however, this brief explanation below will give you a good idea of how I'll read your charts and reports and what can be learned from them.

Noble Sciences Reports

Noble Sciences tools track key aspects of energy movement. I analyze more than 90 charts in a given period and take into account major patterns that appear and disappear over time. When looking at an individual's chart and a chart for a given day, I decode the flow of energy using the Mandala created in a day's conditioning field (six month period of development) for clues and hints about energy patterns that show how you might best orient yourself. Over time major patterns and keys in planetary movements tell a story. I read the astrological meaning of each planet, their relationship to each other, where they are in the body energy maps and what influence they have based on the unique chart, and how they relate to the meaning of the I-Ching Gates.

After absorbing all this and so much more than can be described here, numerous reports offer guidance and support. I provide concrete advice on the best strategy for your mental, spiritual, emotional, and physical well being so you can optimize your daily life with ease. Those who read the Noble Sciences Sacred Synthesis Tools can view the the Daily Action Points at www.noblesciences.com or the Daily Type

Chart (http://www.noblesciences.com/reports/dailytypechart.php) and on Beyond Human Design (http://beyondhumandesign.com/dailytypechart.php) that show the energy of each day. They share with me how they experience major shifts their lives. Here's what a couple of clients have to say about how Noble Sciences has helped them:

> *Thank you very much for putting all the effort in my reading, I very much appreciate it. I continue to review it very carefully. I have been self-studying human design for a while but I have never felt what I feel now when I look at the chart on page 2 "Total Integrated Vehicle-Manifesting Life" because this is what I feel being the total ME. Earlier I always felt that some things were just missing when I looked at my design chart, and reading and learning about centers, and Channel and Gate activations I always felt very strong affinity with some elements that were not defined on my birth moment chart (Gate 12, 1, 8, 2, 14). I was even envious of people having it but now I can see these are parts of me. Your work helped me learn how the access or integrate the different layers in order to live out the fullest expression of my vehicle so to speak. So thank you for giving me this great revelation today and I anticipate continuing to explore my charts and process with you further.*
>
> *Best regards,*
> *Agnes*

> *I wrote Eleanor for a Short Reading to see if it might be a good fit for us to work together. I had no idea what to expect, but her more detailed charts and explanations of my life path were so amazingly accurate to me -- and more hopeful after some deeply troubling life experience -- than other Human Design information I had received. I recognize not that it is possible for me to live the way I am meant to live and thereby be more in the flow with my personal life / energy. Thank you so much, Eleanor!*
>
> *Nancy*

About the Author

Eleanor Haspel-Portner, Ph.D.

Eleanor was born in Brooklyn, New York on December 11, 1944 at 11:10 AM EST. She received her Ph.D. from The University of Chicago, Department on Comparative Human Development, in one of the first interdisciplinary departments in the United States. Eleanor uniquely integrates her background and training in the Social Sciences (psychology, biology, anthropology, sociology) with esoteric studies.

By applying multidimensional **NOBLE SCIENCES TOOLS™** that she developed and validated, Eleanor helps people transform their lives. Throughout her extensive career in private practice as a licensed Clinical Psychologist, Reiki Master/Teacher, and Transformational Life and Relationship Coach, Eleanor helped thousands of individuals, couples, and groups synthesize their life experiences in practical ways for living healthy, successful, and creative lives.

Eleanor strongly believes that each individual's core Self manifests fully when given support and encouragement. She also believes that many people simply need some directional help to feel empowered in their lives. Eleanor and her husband, Marvin, work together in documenting **NOBLE SCIENCES TOOLS™**. They met in India on August 13, 1978 and have been married since then. They live closely together with their dog, cats, and pet ducks. They very much enjoy their children and grandchildren.

Eleanor's first book, "Marriage in Trouble: A Time of Decision" was published in 1976 by Nelson-Hall. In recent years her writing and publishing efforts have been focused on developing Noble Sciences Materials and Tools and self-publishing her articles, books, and graphics.

You can contact Eleanor at: ehp@noblesciences.com or call: (310) 230-7787.

Acknowledgments

After finishing my doctorate in 1971, I began having dreams about knowing with a "K." With my extensive background in dream analysis and Jungian analytic work, I recognized that the dreams were leading me on a quest toward intuitive knowing, and I began wondering what doors would open before me. I followed my inner guidance and changed my life accordingly. The shifts in my life culminated when I began the practice of Transcendental Meditation followed by a significant energy experience that took me through a portal of awareness from which there was no turning back.

Thus, began my quest to learn what I could about consciousness and its dimensionality beyond what I had studied in traditional psychology. I shifted my focus in my work completely, and, with a book contract in hand, wrote my first book, *Marriage in Trouble: A Time of Decision*, a book based on my experiences in my own relationship as well as on my clinical work with clients I had seen in therapy. I continued meditating and exploring esoteric disciplines that could reach the depth of soul of my clients and more effectively explain what I was observing.

One of my first mentors in this quest was a great astrologer, Katherine de Jersey whose astrological reading on me in 1973 exposed me to the extraordinary power of this discipline in the hands of a skilled professional. When I consciously acknowledged that I had seen auras from early childhood, I sought grounded knowledge in how to use these skills in my private clinical practice. I studied many psychics as well as studying Jungian analytical psychology in a Jungian training program, and Reiki healing.

As a twice daily meditator, I trusted my "Knowing" when I was led to travel to India in 1978 where I met my soul mate, Marvin, at Osho's ashram in Poona, at a day and time that had been foretold by my dear friend Katherine de Jersey.

With this extensive training in esoteric disciplines and my strong social sciences research degree from the University of Chicago, I felt empowered to work in my professional field as both a clinician and a researcher. I am deeply grateful for my education and for the many teachers who seeded my consciousness.

The story about my work and exploration in the field of Human Design and its evolution is told in Cosmic Secrets Revealed. Nevertheless, I want to acknowledge those who entrusted their souls to me for readings and to those who have helped the work develop. Without Erik Memmert and his Neutrinos through Windows program none of the calculations for basic charts and for the more extensive multi-dimensional charts would exist in workable form. I am ever grateful to him. In addition, Cindy O'Connor Smith helped me with graphics and with formulating some of the images used in this document in their early iterations, her help was invaluable as has been her support and friendship.

My clients through the years who trust me and encourage me with their respect and admiration give me the courage to trust my own knowing and to pursue it even in the face of challenges and the enormous effort this work requires.

My family has been my laboratory as well as my anchor. Marvin encourages me each step of the way and honors my work in all ways essential. And my children have stood by me supporting my explorations, tolerating my testing of theories on them, and laughing with me at the unending dimensionality that we continually explore. My family has travelled with me to the reaches of the multiverse enjoying the unknown and the unexpected as we together discover dimensions still to be found.

In putting this book together I want to thank Brijit Reed, Jones Pinsker, and Michelle White for their editorial and graphic help in putting Cosmic Secrets together in its current form. I am deeply grateful for their expertise and patience with their patience in handling the complexity and process involved in putting this work together.

I look forward to serving you and sharing my passion and joy for this work with you.

In Loving Light,

Eleanor Haspel-Portner, Ph.D.

Eleanor Haspel-Portner, Ph.D.
Pacific Palisades, California

P.S. If you would like to chat, just give me a call and let me reveal to you some cosmic secrets about your deepest self. Talking lets me know how I may best be of service to you. Call me at 310-230-7787 or email me at ehp@noblesciences.com.

www.ingramcontent.com/pod-product-compliance
Lightning Source LLC
Chambersburg PA
CBHW070117080526
44586CB00013B/1323